FROM 'SCIENCE IN THE MAKING' TO UNDERSTANDING THE NATURE OF SCIENCE

"It is clear the author knows a great deal about the relevant history of 'science in the making' and understands why this knowledge is important to students and teachers of science. The use of actual cases in the history of science to show how disagreements between scientists arise and finally are resolved will help students of science better understand the nature of science."

Ronald G. Good, Louisiana State University, USA

"This useful resource for teacher educators and science education researchers collates in one volume the substance of a large body of the nature of science literature from multiple sources. It offers useful information for students wishing to review some of the important historical experiments pertinent to the science concepts presented in secondary and tertiary textbooks."

Kevin de Berg, Avondale College, Australia

The nature of science is highly topical among science teacher educators and researchers. Increasingly, it is a mandated topic in state curriculum documents. This book draws together recent research on nature of science studies within a historical and philosophical context suitable for students and teacher educators. Traditional science curricula and textbooks present science as a finished product. Taking a different approach, this book provides a glimpse of 'science in the making'—scientific practice imbued with arguments, controversies, and competition among rival theories and explanations. Teaching about 'science in the making' is a rich source of motivating students to engage creatively with the science curriculum.

Readers are introduced to 'science in the making' through discussion and analysis of a wide range of historical episodes from the early 19th to early 21st centuries. Recent cutting-edge research is presented to provide insight into the dynamics of scientific progress. More than 90 studies from major science education journals, related to nature of science are reviewed. A theoretical framework, field tested with in-service science teachers, is developed for moving from 'science in the making' to understanding the nature of science.

Mansoor Niaz is a Professor in the Chemistry Department at Universidad de Oriente, Cumaná, Venezuela.

FROM 'SCIENCE IN THE MAKING' TO UNDERSTANDING THE NATURE OF SCIENCE

An Overview for Science Educators

Mansoor Niaz

Routledge
Taylor & Francis Group
NEW YORK AND LONDON

First published 2012
by Routledge
605 Third Avenue, New York, NY 10017
4 Park Square, Milton Park, Abingdon, Oxon OX14 4RN

Routledge is an imprint of the Taylor & Francis Group, an informa business

Library of Congress Cataloging in Publication Data
Niaz, Mansoor.
From "science in the making" to understanding the nature of science : an overview for science educators / Mansoor Niaz.
p. cm.
1. Science—Study and teaching—Methodology. 2. Science teachers—Training of.
I. Title.
Q181.N77 2011
507.1—dc23
2011025447

ISBN13: 978–0–415–80758–6 (hbk)
ISBN13: 978–0–203–14647–7 (ebk)

Typeset in Bembo
by Swales & Willis Ltd, Exeter, Devon

Dedicated to the fond memories of my father, Niaz-Ud-Din Ahmad, who guided me into the historical labyrinths of Ibn-Khaldun and Toynbee, that intertwined at the same time.

CONTENTS

PREFACE

The rationale of 'science in the making' is based on a history and philosophy of science perspective which involves various interactive processes based on pre-suppositions of the scientist, alternative interpretations of the data, controversies among scientists having similar experimental data, inconsistencies involved in the construction of a theory, and the theory-laden nature of scientific knowledge. The history of science bears witness to these and the essence of science is best characterized by the creativity and imagination of the scientists. In contrast, the traditional science curriculum and textbooks espouse an entirely opposite strategy of presenting science as a finished product, in which students simply regurgitate experimental details. According to some researchers, such presentations constitute a "false" image of science, which is not conducive toward a better understanding of science. This leads to the question: Why do we deny our students an image of science based on how it is practiced by scientists ('science in the making')? Based on a critical analysis of various historical episodes, this book provides plausible answers.

The main objective is to familiarize students, teachers, and researchers with 'science in the making' through various historical episodes, such as: Discovery of the planet Neptune; Discovery of the elementary particle neutrino; Dalton's determination of the law of multiple proportions; Maxwell's kinetic theory of gases; Mendeleev's periodic table; Thomson's discovery of the electron; Rutherford's nuclear atom; Bohr's model of the atom; Millikan's determination of the charge of the electron; Millikan's determination of Planck's constant *h*; Determination of wave-particleduality by de Broglie; and Perl's determination of the Tau Lepton. This provides a rich landscape of scientific endeavor covering a period of over 200 years. Some of the salient features of this book:

a. It shows how 'science in the making' is a rich source of motivating students to engage creatively with the science curriculum.
b. It discusses and critically analyzes a wide range of historical episodes from Dalton (early 19th century) to Perl (early 21st century).
c. It presents recent cutting–edgeresearch that provides insight into the dynamics of scientific progress (based on Nobel Laureate Martin Perl's discovery of the Tau Lepton).
d. It looks at how the views of Nobel Laureate Leon Cooper can influence in–service teachers' understanding of the nature of science.
e. It reviews over 90 studies from major science education journals (2004–2008), related to the nature of science.
f. It offers a theoretical framework developed and field tested with in–service science teachers: Presuppositions, Research questions, Heuristic principles, Designing experiments, and Understanding the nature of science.
g. It includes a new *scenario* in the classroom in which students and teachers could present arguments and counter–argumentsbased on historical reconstructions of various episodes in the history of science.

In writing this book my objective was not any particular course. This has the advantage that the book could be adopted for various types of courses, such as: Teaching the nature of science, Introduction to the history and philosophy of science, Research methodology. The book is rich in content–basedissues related to various historical episodes. The intended audience for this book is secondary and university–levelteachers, science teacher educators, researchers in science education, science methods course teachers, and students.

Chapter 2 establishes a relationship between 'science in the making' and heuristic principles within a historical context. Research relating to students' and teachers' understanding of the nature of science is reviewed in Chapter 3. Next, Chapter 4 explores the difficulties involved in introducing the nature of science to in–servicechemistry teachers. Chapter 5 draws attention to the need for differentiating between experimental data and heuristic principles. How the views of Leon Cooper (Nobel laureate) can influence in–servicescience teachers' understanding of the nature of science is the subject of Chapter 6. Martin Perl's (Nobel laureate) perspective on the nature of science and teaching science is presented in Chapter 7. The contents of this book are organized around three main themes: (a) Chapters 2 and 3 deal with 'science in the making' in a historical context and with students' and teachers' understanding of the nature of science; (b) Chapters 4 and 5 explore the experiences of classroom teachers with respect to heuristic principles and the nature of science; and (c) Chapters 6 and 7 deal with two Nobel laureates' perspectives on the nature of science. Readers can select the chapters that address their particular interests.

ACKNOWLEDGMENTS

Like any intellectual endeavor this book has benefitted from the advice and support of many students, colleagues, and friends. My institution, Universidad de Oriente (Venezuela) has provided support for most of my research activities. Leon Cooper's (Nobel laureate, Brown University), perspective on the nature of science and the integration of history and philosophy of science within science textbooks was a major source of inspiration. Cooper's perspective facilitated the design of a teaching strategy for familiarizing science teachers with the nature of science (Chapter 6). Martin Perl (Nobel laureate, Stanford Linear Accelerator Center) has over the years been supportive of my work and especially with respect to "teaching science as practiced by scientists" (Chapter 7). I have benefitted from discussions and criticisms at different stages from: Fouad Abd-El-Khalick (University of Illinois, Champaign-Urbana), Stephen Klassen (University of Winnipeg), Liberato Cardellini (Università Politecnica delle Marche, Italy), and Michael R. Matthews (University of New South Wales).

I would like to thank the two reviewers who provided constructive criticisms and, at the same time, encouragement for completing the book.

My wife (Magdalina) and daughter (Sabuhi) provided invaluable support and the congenial atmosphere necessary for this project.

Finally, a special word of thanks is due to Naomi Silverman, Publisher (Routledge, New York) who found the project feasible from the beginning and continued to support it throughout the different stages of publication.

1

INTRODUCTION

Research in science education has recognized the importance of history and philosophy of science (HPS) for teaching science. A review of this research shows that most students and teachers do not have adequate epistemological views of the nature of science (NOS). This raises many issues, such as: (a) Why is it important to understand how science works? (b) Is it not sufficient for students to learn the content of science? (c) Do students have to learn how and why a scientist performed an experiment? (d) Does it help students to know that the same experimental data was interpreted differently by another scientist? (e) Do we present a false image of science in our textbooks and classrooms? (f) Is the false image of science conducive towards a better understanding of science? (g) Does the science curriculum motivate students to engage creatively and form part of a responsible citizenry? These issues impinge on the NOS and this book provides plausible answers. Hodson (2009) has emphasized the need for a science curriculum that deals with such NOS issues:

> We should ask why a false or confused NOS knowledge constitutes a major problem for science education. In short, why does it matter what image of science is presented and assimilated? It matters insofar as it influences career choice, and so may have long-term consequences for individuals. It matters if the curriculum image of science is such that it dissuades creative, non-conformist, and politically conscious individuals from choosing to pursue science at an advanced level . . . Failing to provide every student with an adequate understanding of the nature of science runs counter to the demand for an educative citizenry capable of responsible and active participation in a democratic society. *(pp. 142–143)*

Nobel laureate Kenneth G. Wilson[1] has similarly emphasized that history of science, "helps students considering science as a career to think, ask questions, and explore the concepts and ramifications of broad topics, enabling them to grasp what science is about and how it is conducted" (Gooday, Lynch, Wilson, & Barsky, 2008, p. 323).

Science textbooks have generally been found to emphasize the empiricist perspective according to which experimental findings unambiguously lead to the formulation of scientific laws and theories. The present state of our textbooks can be summarized in the following terms: "These trends are incommensurate with the discourse in national and international science education reform movements" (Abd-El-Khalick,Waters, & Le, 2008, p. 835). In a similar vein, Winchester (2006) has cautioned: "However one characteristic stands out for most textbooks in our own time, namely, that they are concentrated presentations of results of previous thought, thought that in fact had a long history. And that history is usually ignored" (p. 1).

This should be cause for concern for most science teachers and especially those interested in the HPS. Such a state of our textbooks is even more troublesome if in retrospect we consider what Holton warned almost four decades ago with respect to what he called the myth of *experimenticism*, namely scientific research as the inexorable result of the pursuit of logically sound conclusions from experimentally indubitable premises:

> Almost every science textbook of necessity places a high value on clear, unambiguous, inductive reasoning. The norm of rationalism in the classroom would seem to be threatened if the text were to allow that a correct inductive generalization may be made without unambiguous experimental evidence. Hence, the likelihood is *a priori* great that any pedagogic presentation of any subject will suggest a clear genetic link from experiment to theory. *(Holton, 1969, p. 974, original italics)*

More recently, historian and philosopher of chemistry Trevor Levere, addressing the 7th International History, Philosophy and Science Teaching Conference, Winnipeg, Canada, expressed a similar concern in cogent terms:

> many authors of science textbooks still write as if there were such a thing as *the* scientific method, and use labels like induction, empiricism, and falsification in simplistic ways that bear little relation to science as it is practiced. *(Levere, 2006, pp. 115–116, original italics)*

'Science as it is practiced' by scientists, as suggested by Levere, can indeed be an important guideline for science textbooks and teaching science. This leads to an intriguing question: Why do we deny our students the dynamics of scientific progress based on science as a human enterprise ('science in the making')? Of course, there is no simple answer to this question. One plausible answer could be that traditional science education does not comprehend the creative and contingent nature of science. Philosopher–physicistJames Cushing (1989) has referred to this in the following thought provoking terms: "science is an historical entity whose practice, methods and goals are *contingent.* There may not be *a* rationality which is the hallmark or the essence of science" (p. 2, original italics. In a footnote Cushing explains what he means by *contingent,* "I simply mean not fixed by logic or necessity", p. 20). This might sound sacrilegious to traditional science teachers and textbook authors. Holton, Levere, and Cushing are trying to present a historical perspective of how science is practiced by scientists, namely construction of a scientific theory involves various interactive processes such as: presuppositions of the scientist, alternative interpretations of data, controversies among scientists having similar experimental data, inconsistencies involved in the construction of a theory, and the theory–ladennature of scientific knowledge.

History of science bears witness to these difficulties and the essence of science is perhaps characterized by the creativity and imagination of scientists. Under this perspective, telling students that scientists are rational would be too simplistic and it would be more motivating to reconstruct the different historical episodes in order to illustrate 'science in the making' and how science is practiced by scientists (Niaz, 2010a). In other words, the construction of knowledge requires assumptions that support reasoning within a social and cultural context (Longino, 1990, p. 219). Discussion of historical episodes can provide students with an opportunity to glimpse the complexity of the scientific enterprise and appreciate how, "both rationality and objectivity come in degrees and that the task of good science is to increase these degrees as far as possible" (Machamer & Wolters, 2004, p. 9).

According to Schwab (1974) within a historical perspective, scientific enquiry is based on a conceptual structure of a discipline:

> The structure of a discipline consists, in part, of the body of imposed conceptions which define the investigated subject matter of that discipline and control its inquiries . . . we cannot, with impunity, teach the conclusions of a discipline as if they were about the whole subject matter and were the whole truth about it. For the intelligent student will discover in time—unless we have thoroughly blinded him by our teaching—that any subject behaves in ways which do not conform to what he has been told about it. *(p. 166)*

Translating this into an HPS context, the structure of a discipline would represent the guiding assumptions, theoretical framework, and presuppositions of the scientist. This helps the scientist to formulate research questions, operationalize heuristic principles, design experiments, and finally interpret the results. This process helps our understanding of the NOS. Actually, Schwab goes beyond by alluding to the changing nature of a discipline and hence the need to make students aware of it. This advice from Cushing (1989), Holton (1969), Levere (2006), and Schwab (1964) has not only been ignored but, rather, most science curricula and textbooks espouse an entirely opposite strategy of presenting science as a finished product (final form, cf. Duschl, 1990) based on a "rhetoric of conclusions," which does not facilitate our understanding of 'science in the making' and, hence, the NOS.

At this stage it is important to note that the NOS is an important area of research and of considerable interest to science educators. In a recent study Chang, Chang and Tseng (2010) analyzed the content of science education research based on the scientometric method of multi–stageclustering. These authors found a total of 3,039 articles from four major science education journals, namely *International Journal of Science Education, Journal of Research in Science Teaching, Research in Science Education*, and *Science Education*, during the period from 1990 to 2007. Multi–stage clustering facilitated the identification of nine important topics (clusters), of which "Conceptual change and concept mapping" had the highest number (n = 553) of articles. The next topic of importance was "Nature of science and socio–scientific issues" with 191 articles, published by authors from various countries around the world. Furthermore, in 1990 there were only two articles on NOS–related issues and this number increased to 25 in 2006. This clearly shows the increasing importance of the NOS for teaching science.

In order to facilitate students' and teachers' understanding of the NOS, it is essential that they are provided with a glimpse of scientific practice imbued with arguments, controversies, and competition among rival theories and explanations (cf. Niaz, Aguilera, Maza, & Liendo, 2002). Based on this perspective, the objective of this book is to explore 'science in the making' in order to understand the NOS and, consequently, to draw conclusions for teaching science. In this chapter I shall use two episodes from the history of science (discovery of the planet Neptune and the elementary particle the neutrino), to illustrate how 'science in the making' can be helpful for understanding the NOS.

Discovery of the Planet Neptune

Discovery of this planet is a good example for illustrating 'science in the making.' Neptune was the first planet to be discovered due to evidence that indicated that it was causing a gravitational effect leading to irregularities in the orbit of another planet, Uranus (discovered in 1781 by Friedrich W. Herschel). Thus, scientists predicted the existence of Neptune before it was observed (Grosser, 1962). In 1845, John C. Adams at St. John's College, Cambridge, estimated the orbit of the unknown planet to be beyond that of Uranus, and predicted that it could account for the irregularities in its motion. Later, Urbain J.J. Leverrier in France made similar calculations in 1846 and communicated them to the French Academy of Sciences and Johann G. Galle in Berlin, who discovered the planet on Septem–

ber 23, 1846. At the time of its discovery Neptune was only 1° from the place predicted by Leverrier and about 2½° from the place predicted by Adams.

Interestingly, Adams had communicated his calculations among others to the English astronomer James Challis at Cambridge. Challis undertook to verify the calculations of Adams and Leverrier, especially with respect to the existence of a new planet (for details, see Smart, 1946). Challis sighted the undiscovered planet (i.e., Neptune) at least four times during the summer of 1846 (once on August 4), that is before Galle. According to philosopher–physicalchemist Michael Polanyi (1964), "these facts made no impression on him [Challis], for he distrusted altogether the hypothesis which he was testing" (p. 30). This clearly shows how lack of a belief in a presupposition (existence of Neptune) led Challis to ignore relevant experimental data.

Now let us see how a physicist–philosopherof science has interpreted the discovery of Neptune based on a conjecture:

> Leverrier and Adams [must have wondered] "Look here, the planet Uranus is not keeping time properly; the only way we can both acknowledge that fact and also save celestial mechanics is to suppose that there is another object, some "dark body," which has the following properties, a, b, c . . . etc." And they worked out the properties of this "in reverse," as it were. What would have to be the properties of a planet in order to perturb Uranus as it is perturbed? *(Hanson, 1964, pp. 166–167)*

This constitutes an interesting example of 'science in the making.' Early conjectures of Leverrier and Adams, subsequent discovery of Neptune by Galle, and the interpretation by Hanson, are all based on the premise that Newton's physics and especially the law of gravitation correctly described the orbits of the planets. Hanson (1958) pays tribute to the intellectual efforts of Leverrier in the following terms: "How remarkable that this man [Leverrier] should have raised classical mechanics to its highest pinnacle by predicting the unseen Neptune as being responsible for observed aberrations in the orbit of Uranus" (pp. 203–204).

Lakatos (1970) goes beyond and provides further insight by presenting an imaginary case of planetary misbehavior that elucidates how scientists do science:

> A physicist of the pre–Einsteinianera takes Newton's mechanics and his law of gravitation (*N*), the accepted initial conditions, *I*, and calculates, with their help, the path of a newly discovered small planet, *p*. But the planet deviates from the calculated path. Does our Newtonian physicist consider that the deviation was forbidden by Newton's theory and therefore that, once established, it refutes the theory *N*? No. He suggests that there must be a hitherto unknown planet *p′* which perturbs the path of *p*. He calculates the mass, orbit, etc., of this hypothetical planet and then asks an experimental astronomer to test this hypothesis. The planet *p′* is so small

> that even the biggest available telescopes cannot possibly observe it: the experimental astronomer applies for a research grant to build yet a bigger one . . . Were the unknown planet p' to be discovered, it would be hailed as a new victory of Newtonian science. But it is not. Does our scientist abandon Newton's theory and his idea of the perturbing planet? No. He suggests that a cloud of cosmic dust hides the planet from us . . . But the cloud is not found. Is this regarded as a refutation of Newtonian science? No . . . [and] yet another ingenious auxiliary hypothesis is proposed . . .. *(pp. 100–101)*

Motterlini (1999, p. 69) considers that the imaginary story of the planet related by Lakatos is based on many real historical instances including the discovery of Neptune. This story encapsulates many aspects of 'science in the making' and thus has implications for understanding the NOS, as follows: (a) When confronted with empirical evidence that seems to refute a scientific theory, scientists generally resist such a refutation and look for an alternative hypothesis; (b) The alternative hypothesis requires further experimental evidence (mass, orbit, and other characteristics of an unknown planet, for example the work of Adams and Leverrier in the case of Neptune); (c) The process of finding alternative hypotheses and looking for experimental support can continue for some time; (d) The role of these "auxiliary hypotheses" is to protect the guiding assumptions or hard–core of a theory (Newtonian theory in the present case); (e) Eventually, the hard–core of a theory crumbles and a new theoretical framework assumes the role of theory building (Einstein's general relativity theory, 1915, in the present case).

Discovery of the Elementary Particle Neutrino

Before 1930 it was generally believed that, based on Einstein's equation, $E = mC^2$, mass–energyis conserved in nuclear reactions. Based on this assumption, generally referred to as "energy conservation" whenever there is a change of mass in nuclear reactions, the difference shows up as kinetic energy, as indicated by Einstein's equation. By the end of the 1920s it was found that energy conservation does not seem to hold for beta decay reactions (changing a neutron into a proton and an electron in radioactivity), as about one–thirdof the energy seems to disappear. To uphold the law of conservation of energy it was postulated that another particle is emitted that carries off the missing energy. This implied the existence of particles called neutrinos, predicted as early as 1929 by W. Pauli, years before they were actually discovered.

Although the neutrino could not be detected for many years, it became important after Enrico Fermi presented his theory of beta decay in 1933 in which neutrinos (Italian for "small neutral one") are emitted and by 1940 it was used routinely by nuclear theorists (Kragh, 1999). Fermi's theory identified the weak nuclear force as being distinct from the strong nuclear force and responsible for beta decay. Interestingly, Fermi's ground breaking theory of beta decay, which founded the modern theory of weak interactions, was first rejected by *Nature*. Neutrinos are massless, chargeless, do not feel the strong nuclear force, and interact via the very short-rangedweak nuclear force. Recent research based on neutrino oscillations, however, has shown that neutrinos might have mass. Actually physicists believed in the existence of the neutrino even though it had not been detected, and for some it was simply a convenient way of organizing experimental data.

Despite the difficulties and a lack of interest in the experimental detection of the neutrino, in 1951 Frederick Reines and Clyde Cowan at Los Alamos started planning experiments. In 1956, using the Savannah River reactor as a neutron source, Reines and Cowan found signals that were considered signs of neutrino-proton reactions (Cowan et al., 1956). Reines shared with Martin Perl the 1995 Nobel Prize for physics (Cowan had died earlier). In his Nobel Prize acceptance speech, thoughtfully entitled, "The neutrino: From poltergeist to particle", Reines referred to the original idea of Pauli in the following terms:

> The neutrino of Wolfgang Pauli was postulated in order to account for an apparent loss of energy-momentumin the process of nuclear beta decay. In his famous 1930 letter to the Tübingen congress, he stated: "I admit that my expedient may seem rather improbable from the first, because if neutrons[2] existed they would have been discovered long since. Nevertheless, nothing ventured nothing gained . . . We should therefore be seriously discussing every path to salvation." *(Reines, 1997, p. 203)*

In June 1956, Reines and Cowan sent a telegram to the man who started it all (Pauli) informing him that they had definitely detected neutrinos from fission fragments by observing inverse beta decay of protons. Pauli's response was prophetic indeed and shows yet another facet of 'science in the making,' "Everything comes to him who knows how to wait, Pauli" (Reproduced in Reines, 1997, p. 215).

According to Hanson (1958), "The neutrino idea, like those of other atomic particles, is a *retroductive conceptual construction* out of what we observe in the large" (p. 124, emphasis added). Considering the immense efforts required to detect the neutrino, Kuhn (1970) concluded: "no experiment can be conceived without some sort of theory, the scientist in crisis will constantly try to generate speculative theories that, if successful, may disclose the road to a new paradigm" (p. 87).

In contrast to the traditional textbook science, these two episodes from 'science in the making' (Neptune and neutrino) clearly show that scientists generally resist the refutation of a theory by putting up alternatives and that besides the experimental apparatus a scientist is almost always accompanied by his presuppositions that provide guidance in the face of difficulties.

Schwab versus Hanson: From Structure of a Discipline to Structure of Scientific Knowledge

In this section I plan to contrast the views of educational philosopher Joseph Schwab with those of physicist–philosopherNorwood Russell Hanson. This is based on an exchange between the two at the Fifth Annual Phi Delta Kappa Symposium on Educational Research, held at the College of Education, University of Illinois, 1964. Schwab was then Professor of Education, Graduate School of Education, University of Chicago and Hanson was Professor of Philosophy at Yale University. There were other participants at the Symposium, familiar to science educators, such as: David Ausubel, Professor of Educational Psychology, University of Illinois; Carl Bereiter, Assistant Professor of Educational Psychology, University of Illinois; Egon Guba, Director, Bureau of Educational Research and Service, Ohio State University; Nathaniel Gage, Professor of Educational Psychology, Stanford University. By any standard, this was a very select group of considerable interest to research in science education and the issues discussed bear witness to the intellectual acumen of the participants.

Schwab (1962, 1974) is well known for his *Structure of a Discipline*, and Hanson (1958) for his *Patterns of Discovery*. In his lecture at the Symposium, Hanson emphasized the difference between the Binomial theorem and a description of physical phenomena in binomial form. In contrast to mathematics, subject matter in physics is not exclusively determined by the postulates and principles of inference. After providing various examples, Hanson (1964) concluded: "*No statement of pure mathematics can be presumed necessarily true when adapted to physical inquiry*" (p. 152, original italics). For example, it is not a mathematical truth that a body will either remain at rest or else move uniformly and rectilinearly to infinity, in the absence of impressed forces. Although this is a standard assumption in kinematics, Hanson wanted students to know that Aristotle and two millennia of Aristotelians would have denied such a claim. After Hanson (1964) finished his lecture, the following exchange took place with Schwab (in order to avoid lengthy sequences, some of the responses have been shortened):

Schwab: . . . an idea developed by Einstein that the greatest mistake that we make about physics is to suppose that it is an "inductive" science. Einstein and Whewell . . . suggest that physics, quite the contrary, is the imaginative invention of an essentially mathematical construction adequate to subsume the data which one is concerned to organize and account for (p. 164).

Hanson: What you say is false (p. 165).

Schwab: Wait a minute. And consequently when the empirical test is made, as you are insisting that it must be made, it is made not of an isolated proposition alone but on the entire corpus (p. 165).

Hanson: That's alright (p. 165).

Schwab: . . . insofar as the whole big corpus of theory can be treated algorithmically then there is a funny way in which physics and mathematics do mix . . . (p. 165).

Hanson: That's what you are suggesting and that's what I am denying . . . even the most "transparent" principle, like the Principle of Conservation of Energy . . . remain empirically vulnerable claims . . . semantically, the pure physics and the pure mathematics are on opposite sides of the logical ravine (pp. 165–166).

Schwab: Nobody in his right mind could argue against your thesis as to which side of the ravine physics is on (p. 166).

Hanson: Then I don't understand what *you* are arguing about (p. 166, original italics).

Schwab: I am not arguing . . . For example, you know very well that one of the particles that [Wolfgang] Pauli invented was invented precisely for the convenience of preserving one of the conservation laws . . . (p. 166).

Hanson: You are really answering my question for me, because it was the nature of that "invention" of the *neutrino* (in 1929 and 1930), an invention generated solely in order to save the conservation principle, which threw a shadow of dubiety on that particular discovery. It wasn't until the empirical work of Cowan and Reines in 1956 and 1957 (at Savannah River) that the neutrino became fully respectable; *there* was an observable effect that showed the physicist not only to be *inventing* entities to save a theory, but also to be discovering empirical evidence for this invention (p. 166, original italics).

Schwab: I agree (p. 166).

Hanson: [Anderson told me]: "I don't believe there is any such thing. All they (Cowan and Reines) show are some numbers, and not all of the numbers. I can just barely believe there is a genuine effect." . . . what Anderson was saying then was this: "If you really want me to entertain the neutrino as a physical entity capable of all the explanatory tricks the theoreticians want, then show me something, in a cloud chamber or somewhere . . . I want to see what the *difference* is that answers to the name 'neutrino.'" Now you are quite right to point out that, to theoreticians, this kind of complaint doesn't mean much—or not very much (p. 169, original italics). [Carl D. Anderson, a former student of Robert Millikan discovered the positron in 1932. Hanson met him at the High Energy Conference in Rochester in 1957, and the remark cited above was made when they were discussing the evidence for the existence of the neutrino].

Schwab: . . . for every ten Andersons there is one Fermi, who said it would be nice if neutrinos were verified in the way which neutrons, protons, and electrons were, but I think it would be helpful to adopt it now (p. 171).

Hanson: Well, that is a nice statement about *you*, Joe . . . All I am trying to call attention to is what you are obscuring (and in so doing you are being "nasty, brutish, and short"). If one stresses what you are stressing, namely, that in physics you . . . (p. 171, original italics).

Schwab: Well, you . . . (p. 171).

Hanson: If I may just finish, *please*. If one stresses what you have been stressing . . . one fails to perceive the fundamental logical difference between every single proposition of physical theory and every single proposition in a purely mathematical algorithm (pp. 171–172, original italics).

At first sight it seems that the issues being discussed by Hanson and Schwab are of minor importance and not of direct relevance to science education. However, I will now elaborate and show that despite the similarities of views the issues being discussed are of fundamental importance for 'science in the making,' understanding the NOS, and teaching science. Furthermore, it is important to note that despite a similar epistemological stance, both Hanson and Schwab make a passionate and rather obdurate defense of their respective positions, leading to some tense moments in the debate. This also shows that understanding the NOS is a difficult enterprise and similar debates have also been observed at science education conferences (for details, see Niaz, 2008a, pp. 135–136).

It seems that the difference between the epistemological positions of Hanson and Schwab can be summarized in the following terms: For Hanson, despite the similarities mathematical propositions are axioms which cannot be adapted in the context of physical science. On the contrary, Schwab would suggest that propositions and their meaning (or premises) do not directly refer to empirical facts, and thus there is something strangely mathematical about physics. Schwab (1964) presents his perspective of scientific knowledge in cogent terms:

> In general, then, enquiry has its origin in a *conceptual structure*. This structure determines what *questions* we shall ask in our enquiry; the questions determine what data we wish; our *wishes* in this respect determine what *experiments* we perform. Further, the *data*, once assembled, are given their *meaning and interpretation* in the light of the conception which initiated the enquiry. *(p. 9, emphasis added)*

Indeed, this constitutes an outline of a research methodology based on: conceptual structure → questions → wishes (i.e., presuppositions) → experiments → data → understanding based on meaning and interpretation. These are important issues for understanding 'science in the making' within an HPS perspective.

At this stage it would be interesting to see how Duhem (1914) an important philosopher of science would view this dilemma: "What the physicist states as the result of an experiment is not the recital of observed facts, but the interpretation and the transposing of these facts into the *ideal, abstract, symbolic world* created by the theories he regards as established" (p. 159, italics added). In the case of a clash between the two (empirical facts and theory), Duhem suggested upholding the experimental facts and considered the theory to be a "parasite," which in a way contradicted his own philosophical position. Now it is plausible to suggest that the *ideal, abstract, symbolic world* comes quite close to what Schwab referred to as propositions and their meaning. The controversy between Hanson and Schwab also reflects a contradiction similar to that of Duhem. Hanson seems to be upholding a position, quite similar to that of Duhem, that considers the experimental facts to be paramount. Schwab, on the contrary, espouses a philosophical position that comes quite close to what scientists do (based on a pluralistic model) under such circumstances and fully endorsed by Lakatos:

> In the pluralistic model the clash is not "between theories and facts" but between two high-leveltheories: between an *interpretative theory* to provide the facts and an *explanatory theory* to explain them; and the interpretative theory may be on quite as high a level as the explanatory theory . . . *the problem is which theory to consider as the interpretative one which provides the "hard" facts and which the explanatory one which "tentatively" explains them.* In a mono-theoreticalmodel we regard the higher-leveltheory as an *explanatory theory to be judged by the "facts"* delivered from outside (by the authoritative experimentalist): in the case of a clash we reject the explanation. *(Lakatos, 1970, p. 129, original italics)*

According to Lakatos, based on a mono-theoreticalmodel a theory can be rejected on the sole grounds of experimental evidence. 'Science in the making' (based on the pluralistic model) shows that rejection of a theory is not a simple and straightforward question of accepting or rejecting experimental evidence. On the contrary, interpretation of experimental evidence is extremely difficult, which leads to conflicts and controversies among contending groups of scientists. Within the Lakatosian framework the "hard-core"(negative heuristic) of a research program is resistant to refutation and may even be based on contradictory and inconsistent foundations (for details, see Niaz, 2011, pp. 15–16). Similarly, Giere (2006) has endorsed the pluralistic view of progress in science based on "perspectival realism" (p. 5). In contrast to "objective realism," Giere espouses a perspective according to which no theory can provide us with a complete and literally correct picture of the world.

History of science shows that scientific controversies at times can continue for decades and are generally brought to a closure by the intervention of the scientific community (e.g., Millikan's oil drop experiment, cf. Niaz, 2005a). It is precisely in this respect that the Lakatosian framework goes beyond Duhem by suggesting that scientists are guided by their presuppositions (hard-coreof beliefs) and they resist any change in this *ideal, abstract* and *symbolic* world. Niaz (2009a, Chapter 3) has presented a detailed comparison of the philosophies of Duhem and Lakatos and concluded that scientific endeavor depends to a large degree upon the imagination and creativity of the scientists.

In this context, it is now possible to understand better the debate between Hanson and Schwab, especially with respect to the discovery of the neutrino. According to Schwab, scientific propositions (presuppositions, Holton, 1978; hard-core of beliefs, Lakatos, 1970) may be accepted even before confirming experimental evidence becomes available. Two leading physicists-historians of science would endorse a similar thesis in categorical terms: "Yet physicists had so much faith in the law of conservation of energy that they preferred to believe in an apparently unobservable particle [neutrino, suggested by Pauli] rather than abandon the law" (Holton & Brush, 2001, p. 502).

Similarly, Lawson (2010) has emphasized the role of theory-driven research for science education. 'Science in the making' provides many examples of how it is the theory (presuppositions) that decides what can be considered as data: (a) J.J. Thomson's rejection of E. Rutherford's hypothesis of compound scattering (alpha particle experiments) as he strongly believed in the uniform distribution of mass and charge in his atomic model (plum-pudding); (b) Millikan's determination of the elementary electrical charge based on his presupposition of the atomic nature of electricity; (c) Millikan's acceptance of Einstein's equation to determine Planck's constant h (photoelectric effect) and yet he still rejected the hypothesis of light quanta, as he strongly believed in the classical wave theory of light; (d) controversial experimental evidence of the bending of light in the 1919 eclipse experiments to support Einstein's general theory of relativity (cf. Niaz, 2009a, Chapter 9); (e) De Broglie's postulation of wave-particle duality before there was any experimental evidence. According to Schwab (1974) besides the presuppositions, scientific inquiry tends to look for patterns of change and relationships, which constitute the heuristic principles of scientific knowledge. It is precisely these heuristic principles that guide us to look for facts and what meaning to assign these facts. Various historical episodes discussed in Chapter 2, illustrate how the heuristic principles facilitate the designing of experiments.

Chapter Outlines

'Science in the Making' and Heuristic Principles in a Historical Context (Chapter 2). This chapter analyzes various episodes in the history of science based on the following framework: (1) Elaboration of a theoretical framework based on presuppositions; (2) Formulation of research questions; (3) Operationalizing heuristic principles; (4) Designing experiments; and (5) Understanding the NOS. The following episodes that constitute important examples of 'science in the making' were analyzed: (a) Dalton's determination of the law of multiple proportions in chemistry; (b) Maxwell's kinetic theory of gases; (c) Mendeleev's periodic table of chemical elements; (d) Thomson's determination of the mass to charge ratio of cathode rays; (e) Rutherford's alpha particle experiments and the

nuclear atom; (f) Bohr's model of the atom; (g) Millikan's determination of the elementary electrical charge; (h) Millikan's determination of Planck's constant *h*; and (i) Determination of wave–particleduality by de Broglie. After having shown how 'science in the making' in a historical context facilitates an under-standing of the NOS, it is suggested that the next step would be to incorporate these historical episodes in the context of the science curriculum and elaborate appropriate science stories.

Students' and Teachers' Understanding of the Nature of Science (Chapter 3). This chap-ter reviews research on the following aspects and draws implications for science education: (a) Epistemological beliefs of students and teachers with respect to the NOS; and (b) Facilitating students' and teachers' understanding of the HPS, based on topics that are already in the science curriculum. This thematic review focuses on studies published in the period, 2004–2008 and draws upon articles in the following journals: *International Journal of Science Education* (n = 34), *Journal of Research in Science Teaching* (n = 28), and *Science Education* (n = 32). Of the 94 studies reviewed, 60 (65%) are classified in the section on epistemological beliefs. Based on the subject, treatment, and orientation of the study, the following seven categories are generated: (1) Relationship between students' and teachers' epis-temological beliefs (n = 27); (2) Myth of the scientific method (n = 3); (3) Chil-dren's scientific reasoning (n = 4); (4) Scientists' views of the NOS (n = 9); (5) the NOS and the science curriculum (n = 10); (6) the NOS and students' labora-tory practice (n = 6); and (7) Science exhibitions for understanding the NOS (n = 1). Thirty–fourstudies are classified in the section on facilitating students' and teachers' understanding of the HPS, and the following six categories are gener-ated: (1) The role of argumentation (n = 9); (2) Explicit and reflective vs. implicit inquiry–orientedinstruction (n = 11); (3) The use of NOS–enrichedmaterials (n = 7); (4) The use of history–basedinstructional material (n = 3); (5) The use of technology–basedhistorical materials (n = 2); and (6) The use of science appren-ticeship programs (n = 2).

How to Introduce the Nature of Science in the Classroom (Chapter 4). The objective of this study is to facilitate chemistry teachers' understanding of the NOS and explore difficulties involved in its implementation in the classroom. The study is based on the responses of 16 in–serviceteachers who had registered for an 11-week course on the "Epistemology of Science Teaching," as part of their Master's degree program in education. The course is based on 13 readings drawing on the NOS, critical evaluation of the NOS, and critical evaluation of constructivism. Course activities included written reports, classroom discussions based on partici-pants' presentations, and written exams.

The Role of Heuristic Principles in Understanding the Nature of Science (Chapter 5). Research in science education has drawn attention to the need for differentiating between experimental data and "heuristic principles" that facilitate understanding of the NOS. The objective of this study was to facilitate chemistry teachers' understanding that emphasis on experimental data leads to a "rhetoric of conclusions" and does not facilitate understanding of the NOS. The study is based on 26 in-serviceteachers who had registered for a 10-weekcourse on "Investigation in the Teaching of Chemistry," as part of their Master's degree program. The course is based on 18 readings drawing on the HPS, students' alternative conceptions, and conceptual change. Course activities included written reports, classroom discussions based on participants' presentations, and written exams.

How the Views of Leon Cooper (Nobel Laureate) can Influence In-service Teachers' Understanding of the Nature of Science (Chapter 6). Research in science education has recognized the importance of the NOS for understanding science. Leon Cooper (Nobel laureate, physics, 1972), has presented a framework based on the HPS to facilitate a better appreciation of the dynamics of scientific progress. The objective of this study is to evaluate how the views of a Nobel laureate can influence in-serviceteachers' understanding of the NOS based on a reflective and explicit, activity-basedapproach. The study is based on the responses of 20 participants who had registered for an introductory course as part of their doctoral program. Besides other material, the framework developed by Cooper (Niaz, Klassen, McMillan, & Metz, 2010b) was required reading. The importance of understanding experiments (oil drop, cathode rays, alpha particles, photoelectric, etc.) within an HPS perspective was explicitly discussed in class. At the end of the course all participants were evaluated on the responses to a five-itemquestionnaire, based on assertions derived from Cooper's framework. Participants were required to respond by indicating if they were: (a) In agreement, (b) In partial agreement, or (c) In disagreement, and explain their response.

Martin Perl's (Nobel Laureate) Perspective on the Nature of Science and Teaching Science (Chapter 7). Martin L. Perl was the recipient of the 1995 Nobel Prize in physics for his discovery of the Tau Lepton, based on a 16-yearhistory (1963–1979), when all experimental measurements agreed with the hypothesis that the Tau was a lepton produced by a known electromagnetic interaction. Besides this, Perl has also worked on the isolation of elementary particles with fractional electric charge, namely quarks. Based on his experience as an experimental scientist, Perl has formulated a philosophy of speculative experiments (Perl, 2004; Perl & Lee, 1997). The objective of this chapter is to present a brief account of the discovery of the Tau Lepton and work on quarks, in order to understand the NOS and then draw implications for teaching science.

Nature of science manifests itself in the different topics of the science curriculum as heuristic principles. Textbooks, by emphasizing not only the empirical nature of science but also the underlying heuristic principles, can be particularly helpful in facilitating conceptual change (Niaz, 2001a). It is plausible to suggest that 'science in the making' based on historical reconstructions can provide students and teachers with innovative teaching strategies in order to facilitate a better understanding of the nature of science (Niaz, 2011).

Notes

1 Kenneth G. Wilson was awarded the 1982 Nobel Prize in Physics for his theory of critical phenomena in connection with phase transitions. In Gooday et al. (2008), Wilson has posed an interesting question: "Does science education need the history of science?" and responded in the affirmative by suggesting that the history of science be included in the science curriculum.

2 When the neutron was discovered by Chadwick in 1932, Fermi renamed Pauli's particle the 'neutrino'.

2

'SCIENCE IN THE MAKING' AND HEURISTIC PRINCIPLES IN A HISTORICAL CONTEXT

According to Schwab (1974) scientific research has its origins not in objective facts alone but in a deliberate construction of the mind—heuristic principle, which guides the scientist to look for facts and interpret the findings. Based on the existing knowledge in a field of research a scientist formulates the guiding assumptions (presuppositions) of the research program. One of the most important characteristics of the presuppositions is that scientists generally do not abandon them in the face of anomalous data (Niaz, 2009a). The role of presuppositions in scientific progress has been recognized by philosophers of science in the following terms: "hard-core"of a research program (Lakatos, 1970), guiding assumptions (Laudan, Laudan, & Donovan, 1988), and presuppositions (Holton, 1978). Based on history and philosophy of science (HPS) the objective of this chapter is to show that the presuppositions lead to research questions which are then operationalized through heuristic principles (Schwab, 1962, 1974), finally leading to the design of experiments, and consequently an understanding of the nature of science (NOS). It is plausible to suggest that scientific progress is a dynamic process based on the following interactive processes:

1. Elaboration of a theoretical framework based on presuppositions, guiding assumptions, hard-coreof beliefs, and previous experience
2. Formulation of research questions
3. Operationalizing heuristic principles
4. Designing experiments
5. Understanding the nature of science

In order to facilitate an understanding of the interactive dynamic process underlying progress in science, various historical episodes are presented in this chapter.

Historical reconstruction of these episodes presents the essential elements and for further details the reader is referred to Niaz (2009a). The historical episodes presented cover a period of almost 200 years from Dalton to de Broglie, and are followed in each case by the five interactive steps outlined above, which have educational implications. A major objective of the historical context is to facilitate not only a better conceptual understanding but also a more humanistic appreciation of the scientific endeavor (Niaz, 2009a; Stinner, 1995). At this stage it is important to point out that the possibility of an alternative process from the one presented here, leading to scientific development is not ruled out. It is plausible to suggest that the historical context is essential for introducing the NOS in both the classroom and textbooks (Winchester, 2006). The following episodes from the history of science provide an illustration of this approach:

a. Dalton's determination of the law of multiple proportions in chemistry (starting 1800).
b. Maxwell's kinetic theory of gases (starting 1860).
c. Mendeleev's periodic table of chemical elements (starting 1869).
d. Thomson's determination of the mass to charge ratio of cathode rays (starting 1897).
e. Rutherford's alpha particle experiments and the nuclear atom (starting 1911).
f. Bohr's model of the atom (starting 1913).
g. Millikan's determination of the elementary electrical charge (starting 1910).
h. Millikan's determination of Planck's constant *h* (photoelectric effect, starting 1913).
i. Determination of wave–particleduality by de Broglie (starting 1924).

It is important to note that most of these historical episodes constituted crucial steps in scientific progress during the twentieth century. In order to facilitate a better understanding of these episodes I have used material from my previous publications, including Niaz (2009a, 2011).

Dalton's Determination of the Law of Multiple Proportions in Chemistry

The French chemist Joseph Louis Proust (1754–1826) was the first to express and provide experimental verification of the law of definite proportions in 1799. The law is generally enunciated in textbooks in the following form: "Different samples of a substance contain its elementary constituents (elements) in the same proportions" (Pauling, 1964, p. 25). Another French chemist Claude Berthollet (1749–1822) was quick to oppose the law by arguing that there was no essential difference between chemical compounds and solutions, which showed variable composition. There ensued an eight–yearcontroversy between Proust and

Berthollet, which ended with the law of definite proportions being firmly upheld (cf. Fujii 1986; Ihde 1964).

The law of multiple proportions is generally attributed to the English chemist John Dalton (1766–1844), who first discovered it in August 1803, according to textbook accounts, while working on the composition of the hydrocarbons methane and ethene. Most textbooks state the law in the following terms: "When an element combines with another to form more than one compound the masses of the second element combining with a fixed mass of the first element bear a simple ratio to one another" (Taylor, 1942, p. 2). For example, the series of oxides formed by nitrogen and oxygen (N_2O, NO, and NO_2) is considered to be a good illustration of the law. Some scholars (Thomson, 1825) in the early nineteenth century popularized the positivist version that Dalton was led to his atomic theory by the discovery of the law of multiple proportions while working on the two hydrocarbons (methane and ethene). According to Rocke (1984): "This inductivist version was quite concordant with the then prevalent Victorian model of heroic science" (p. 27). Linus Pauling (1964) clarifies the issue by stating categorically: "The discovery of the law of simple multiple proportions was the first great success of Dalton's atomic theory. This law was not induced from experimental results, but was derived from the theory, and then tested by experiments" (p. 26).

Most chemistry textbooks present the laws of definite and multiple proportions in a simple and uncontroversial manner. In spite of such presentations, a critical appraisal of the HPS literature shows that the development of the two laws is closely related to the controversial origin of Dalton's atomic theory, starting from the early nineteenth century (cf. Christie, 1994; Frické, 1976; Fujii, 1986; Ihde, 1964; Moore & Hall, 1939; Nash, 1956; Rocke, 1978, 1984; Russell, 1988). For example, the laws of definite and multiple proportions were interpreted by some scientists as a manifestation of Gay-Lussac'slaw of combining volumes. On the other hand, an alternative interpretation considered the laws to be a manifestation of the atomic nature of matter. These opposing interpretations continued almost to the first decade of the twentieth century (cf. Pauling, 1952, pp. 135–136).

The origin of Dalton's atomic research program can be traced to 1801 when he formulated his *first theory of mixed gases*. Frické (1976) considers the *hard-core* of Dalton's atomic research program to be based on the following assumptions:

> that each chemical element consists of its own characteristic type of indivisible, indestructible atom, that the property which characterizes these atoms is their weight, that compounds are definite combinations of these different atoms, and that, during chemical reactions between compounds, atoms are neither created nor destroyed but only rearranged to form new compounds. *(p. 283)*

Dalton's atomic research program provided a rationale for Proust's law of constant composition, explained Richter's law of equivalents, and predicted the law of multiple proportions by weight (cf. Frické, 1976, p. 283; Pauling, 1964, p. 26). Interestingly, just as Dalton was working out the details of his atomic theory, Gay–Lussac(1808) presented his law of combining volumes, which stated: "The volumes of gases which react with one another or are produced in a chemical reaction are in the ratios of small integers" (Pauling, 1952, p. 161). Dalton rejected Gay–Lussac'slaw on the grounds that it seemed to contradict his atomic research program. For Dalton to have accepted Gay–Lussac's law would have amounted to the recognition of the experimental finding that gases combine in simple ratios with respect to their volumes, and thus ignore the explanation based on atomic theory. Gay–Lussac'sown reaction was somewhat ambivalent, as his work was supported by, "his antiatomist mentor, patron, and friend Berthollet" (Rocke, 1984, p. 41). According to Frické (1976):

> Gay–Lussacwas not alone in rejecting atomism. There was a widespread tendency to replace the theoretical concept of "atom" by the *measurable notions* of "volume", "equivalent", or "measure". Dalton's empirical laws, such as the law of multiple proportions, were considered to be of great scientific value, but his *theories were discarded as speculations. (p. 285, original italics)*

It appears that Gay–Lussac's work was conducted very much in the positivist tradition of the early nineteenth century, which emphasized the visible and measurable (volumes of gases) aspects of chemistry to the detriment of the invisible/hypothetical (atoms/molecules) entities. This background explains Dalton's strong reaction against Gay–Lussac'slaw of combining volumes. Dalton's atomic research program (hard–core,see above) in order to be operationalized required the following items as part of the Lakatosian positive heuristic (cf. Frické, 1976, p. 285):

1. Chemical formulas to represent the number and type of atoms in a compound.
2. Atomic weights (masses) to measure the relative mass of an atom.
3. Composition by weight of the compound, namely, gravimetric combining proportions.

Early in Dalton's career only the third item was known and hence the positivist approach precisely concluded: "the superfluous atomic weights and formulas should be replaced by combining 'equivalents', 'measures', or whatever" (Frické, 1976, p. 285). It is plausible to suggest that the law of definite proportions is basically an elaboration of the third item of Dalton's positive heuristic. Furthermore, Gay–Lussac'slaw of combining volumes provided the antiatomists with a rationale for accepting the laws of definite and multiple proportions, without the "superfluous" atomic theory of Dalton. At this stage it is interesting to note that textbooks generally ignore the fact that, when the laws of definite and multiple proportions were formulated, chemists did not know how to calculate chemical formulas and atomic masses of the elements (first two items of the positive heuristic of Dalton's research program).

Based on the historical reconstruction of the laws of definite and multiple proportions it is plausible to elaborate the following aspects of Dalton's research program, namely, 'science in the making':

Dalton's theoretical framework (Presupposition): Atoms of each chemical element have definite properties which are characteristic of the element. Atoms were distinguished quantitatively by the weight of their atoms. Chemical compounds are definite combinations of these atoms.

Research question: As the law of definite proportions in which chemical elements combine had already been established by Proust, it is plausible to ask if the same elements can combine in varying proportions to form a series of compounds.

Operationalizing heuristic principles: Based on the atomic theory, chemical elements were distinguished by the weight of their atoms. Consequently, the possibility of the same two elements combining in different proportions could be explored.

Designing experiments: Dalton carried out a series of experiments to show that nitrogen and oxygen can combine to form various oxides, such as: N_2O, NO, NO_2. According to Levere (2006): "The oxides of nitrogen were star performers in the instantiation of the law of multiple proportions" (p. 122).

Understanding the nature of science: Of the 27 general chemistry textbooks analyzed (published in the USA, Niaz, 2001b), almost all ignored the underlying issues, namely, the law of multiple proportions was derived from Dalton's atomic theory and then tested by experiments. Furthermore, the textbooks confused the issues by asserting that: (a) Gay–Lussac'slaw of combining volumes provided a rationale for accepting the laws of definite and multiple proportions without the atomic theory of Dalton; and (b) Dalton had not understood Gay–Lussac'sand Avogadro's laws as he did not accept the existence of diatomic molecules. Based on the historical reconstruction and analysis of textbooks it is plausible to suggest that some laws in chemistry were derived empirically (e.g., Law of definite proportions). However, in many other cases (e.g. law of multiple proportions), presuppositions, research questions, and heuristic principles, along with the scientists' imagination and creativity, played a crucial role. In other words, what we need is not a simple description of the experimental findings but, rather, an epistemology of science (Matthews, 1994a).

Maxwell's Kinetic Theory of Gases

Formulation of the kinetic molecular theory of gases based on the contributions of various scientists (Bernouilli, Herapath, Joule, Krönig, Clausius, Maxwell, Boltzmann, and van der Waals) is recognized as one of the greatest achievements of nineteenth–centuryscience. A major problem facing these scientists was to go beyond the well–knownmacroscopic hydrodynamical laws (Boyle, Charles, Gay–Lussac, etc.) and present a theoretical framework within which experimental data could be understood as intelligible. In an attempt to understand the relationship between the observable properties (pressure, volume, temperature) of gases and a plausible model for explaining these properties, James Clerk Maxwell (1860) presented the following simplifying (basic) assumptions:

1. Gases are composed of minute particles in rapid motion.
2. Particles are perfectly elastic spheres.
3. Particles act on each other only during impact.
4. Motion of the particles is subject to the mechanical principles of Newtonian mechanics.
5. Velocity of the particles increases with the temperature of the gas.
6. Particles move with uniform velocity in straight lines striking against the sides of the container, producing pressure.
7. Derivation of the distribution law assumes that the x–,y–,and z–components of velocity are independent.

A student may ask: How did Maxwell arrive at these assumptions? According to Achinstein (1991): "They [assumptions] are highly speculative, involving as they do the postulation of unobserved particles exhibiting unobserved motion" (p. 410). Did Maxwell have an independent warrant (i.e., plausibility of the hypotheses) for his simplifying assumptions? It is plausible to suggest that Maxwell's assumptions helped him to progress from simple to complex models of the gases (referred to as *ceteris paribus* clauses by philosophers of science, Lakatos, 1970, p. 102).

Among other assumptions, Maxwell's (1860) paper was based on "strict mechanical principles" derived from Newtonian mechanics and yet at least two of Maxwell's simplifying assumptions (referring to the movement of particles and the consequent generation of pressure) were in contradiction to Newton's hypothesis explaining the gas laws based on repulsive forces between particles. Newton provided one of the first explanations of Boyle's law in his *Principia* (1687) in the following terms: "If a gas is composed of particles that exert repulsive forces on their neighbors, the magnitude of the force being inversely as the distance, then the pressure will be inversely as the volume" (Brush, 1976, p. 13). Apparently, due to Newton's vast authority, Maxwell even in his 1875 paper, "On the dynamical evidence of the molecular constitution of bodies" reiterated that Newtonian principles were applicable to unobservable parts of bodies (cf. Achinstein, 1987, p. 418). Brush (1976) has pointed out the contradiction explicitly: "Newton's laws of mechanics were ultimately the basis of the kinetic theory of gases, though this theory had to compete with the repulsive theory attributed to Newton" (p. 14). Maxwell's research program is yet another example of a program progressing on inconsistent foundations (similar to Bohr, cf., Lakatos, 1970, p. 142). Indeed Clark (1976) considers the seven simplifying assumptions (presented above) as the hard–core(negative heuristic) of Maxwell's research program.

Based on the historical reconstruction of the kinetic molecular theory of gases it is plausible to elaborate the following aspects of Maxwell's research program, namely 'science in the making':

Maxwell's theoretical framework (presuppositions): The seven simplifying assumptions (presented above) constitute Maxwell's presuppositions for which he did not have an explicit warrant based on experimental evidence. These assumptions guided Maxwell's research program and this precisely is the function of such presuppositions in the early stages of research.

Research questions: (a) What is the nature of the elementary particles that constitute gases? (b) Are all interactions among the particles subject to the laws of mechanics? (c) Can the simplifying assumptions be eliminated (in the long run) so as to approach the conditions of "real" gases? (d) Can the assumptions provide an explanation for the observable hydrodynamic properties of gases?

Operationalizing heuristic principles: Intermolecular forces which account for the cohesion of the liquid phase represented a property of the kinetic molecular model, and hence their effect should still be appreciable in the gaseous phase. (This was explicitly explored by van der Waals' contribution in 1873, by postulating a new equation of state.)

Designing experiments: Using his equation of state, van der Waals reproduced Andrews' (1869) experimental results (isotherms for carbon dioxide), which demonstrated the continuity of the transition from the gaseous to the liquid state. Clark (1976) interprets van der Waals' contribution in the following terms:

> *What is important is that the novel predictions were found to be in good agreement with experiment. Application of the heuristic of the kinetic programme had resulted in empirical growth*, in this case the discovery of a *new general law. (p. 60, original italics)*

Understanding the nature of science: In a study based on 14 secondary school chemistry textbooks, de Berg (1989) reported that the history of science receives only scant treatment and, what is more serious, little emphasis is placed on qualitative (i.e., conceptual) understanding of gas properties. Of the 80 exercises on the pressure volume relationship of gases, only five were based on qualitative properties. Finally, De Berg (1989) concluded: "if textbooks are a guide to what students learn in classrooms, it is no wonder that the 'algorithmic' mode of problem solving in gas laws predominates" (p. 119). Of the 22 general chemistry textbooks analyzed by Niaz (2000a) only three presented Maxwell's guiding assumptions (presuppositions) that facilitated an understanding of the NOS and following is an example:

> At this point we want to build a model (theory) to explain *why* a gas behaves as it does . . . laws do not tell us *why* nature behaves the way it does. Scientists try to answer this question by constructing theories (building models). The models in chemistry are speculations about how individual atoms or molecules . . . cause the behavior of macroscopic systems (collections of atoms and molecules in large enough numbers so that we can observe them). A relatively simple model that attempts to explain the behavior of an ideal gas is the *kinetic molecular theory.* This model is based on speculations about the behavior of the individual particles (atoms or molecules) in a gas. *(Zumdahl, 1990, pp. 434–435)*

This presentation emphasizes speculative models and that models develop (tentativeness) in order to explain the behavior of gases. Most teachers would agree that it constitutes a logical, understandable and helpful preamble before presenting the postulates of the kinetic theory. Rodríguez and Niaz (2004a) have reported that of the 30 general physics textbooks analyzed only two presented satisfactorily the role played by simplifying assumptions.

All the general chemistry textbooks ignored that Maxwell's research program, although successful, was also based on an inconsistent foundation (Niaz, 2000a) and following is an example: "From our knowledge of the gas laws and *Newtonian mechanics*, we can try to understand the bulk properties of a gas, such as pressure and temperature, in terms of the behavior of individual molecules" (Chang, 1981, p. 113, emphasis added). Many textbooks explicitly invoke Newtonian mechanics along with Maxwell's presentation of the kinetic theory, without realizing an inherent contradiction. Similarly, Rodríguez and Niaz (2004a) have reported that none of the general physics textbooks made a satisfactory presentation. It is important to note that most textbooks not only ignore the fact that scientists face difficulties and manifest contradictions but, rather, they present the development of the kinetic theory as a simple accumulation of a series of theories. No wonder, students are at a loss to understand why we need yet another theory. According to Méheut (1997) scientists develop models as cognitive implements that unify descriptions and help to predict and explain physical phenomena. It is plausible to suggest the following implications for understanding the NOS: (a) Based on their theoretical framework scientists formulate presuppositions that have some speculative aspects and in general provide guidelines for developing the research programs; (b) Due to the complexity and difficulties involved some aspects of the presuppositions are based on inconsistent foundations. As the research program progresses, empirical growth provides a greater understanding and inconsistencies are progressively decreased.

Mendeleev's Periodic Table of Chemical Elements

The Periodic Table of chemical elements is considered to be an important concept for chemists, teachers, and historians. Besides classification of elements it has predicted new elements and facilitated conceptual understanding. Most historians and philosophers of science have considered it to be a classification, system, table, law and, very rarely, as a theory. Given the long history of the periodic table, it is not surprising that historians and philosophers of science are far from reaching a consensus with respect to its origin, nature, and development (see Niaz, Rodríguez, & Brito, 2004; van Spronsen, 1969, for details).

Most historians consider the International Congress held in Karlsruhe (3–5 September, 1860) as crucial to the development of chemistry and the periodic table in particular. A circular (dated 10 July, 1860) sent by the organizers of the Congress to most outstanding chemists of Europe, outlined its objective as the need to reach a consensus on "More precise definitions of the concepts of atom, molecule, equivalent, atomicity, alkalinity, etc.; discussion on the true equivalents of bodies and their formulas; initiation of a plan for a rational nomenclature" (reproduced in De Milt, 1951, p. 421). Mendeleev attended the Congress and was greatly impressed by Cannizzaro's contribution and in a letter dated 7 September, 1860 summarized an important achievement of the Congress:

> It is decided to take a different understanding of molecules and atoms, considering as a molecule the amount of a substance entering a reaction and determining physical properties, and considering as an atom the smallest amount of a substance included in a molecule. Further, it reached an understanding about equivalents, considered as empirical, not depending on the understanding about atoms and molecules. *(Reproduced in De Milt, 1951, p. 422)*

This clearly shows the tension reflected for many years in Mendeleev's later writings, namely, on the one hand he wanted to understand the nature of atoms and molecules (that is, atomic theory), and on the other he espoused the use of equivalent weights based on empirical findings. Gordin (2002) also recognizes the importance of the Karlsruhe Congress in the development of Mendeleev's thought on the periodic system. The development of the periodic table is closely related to the atomic theory, the role of empirical evidence in the development of theories, and whether the periodic table itself was a simple classification/codification scheme, empirical law, or a theory.

Periodicity in the Periodic Table as a Function of the Atomic Theory

Many students must have wondered how a simple arrangement of the chemical elements could provide predictive and explanatory (accommodation) power to Mendeleev's periodic table? Many chemistry textbooks (Brito et al., 2005) and teachers even suggest and give the impression that the problems associated with the periodic table could not be solved until Moseley (1913, 1914) and others showed that elements need to be ordered according to their atomic number. According to van Spronsen (1969):

> The actual development of the periodic system seemed to require a catalyst! We think it proper to attribute this catalytic action to Cannizzaro's famous Karlsruhe lecture at the 1860 Congress. He *made the distinction between atoms and molecules and defined such concepts as valence*; in our opinion this initiated the second stage of the discovery and started the history proper of the periodic system of chemical elements. *(p. 1, emphasis added)*

In spite of this fairly categorical statement with respect to the role played by the atomic theory by a major historian of the periodic table, many historians attribute its success primarily to empirically observed properties of the elements (inductive generalization). It is important to recall that most of the pioneering work of Mendeleev was conducted from 1869 to 1889, before Thomson (1897), Rutherford (1911), Bohr (1913), and Moseley (1913, 1914) laid the foundations of the modern atomic theory. So how could Mendeleev conceptualize periodicity as a function of the atomic theory? An answer to this question will precisely show Mendeleev's ingenuity, farsightedness, creativity, and the ability to "speculate," and hence 'science in the making.' Despite Mendeleev's own ambivalence and ambiguity, an historical reconstruction does provide a convincing story of this remarkable contribution to our knowledge. Before starting work on the periodic table Mendeleev had the following important sources of information: Dalton's atomic theory, law of multiple proportions, Cannizzaro's Karlsruhe lecture, fairly reliable atomic weights, atomicity (valence), and various physical and chemical properties of the elements. The following examples from Mendeleev's writings provide support for the role played by the atomic theory in establishing periodicity:

a. Even in his first publication Mendeleev referred to the relationship, albeit implicitly, between periodicity, atomic weights, and valence: "The arrangement according to atomic weight corresponds to the *valence* of the element and to a certain extent the difference in chemical behavior, for example Li, Be, B, C, N, O, F" (Mendeleev, 1869, p. 405, original emphasis).
b. In his famous Faraday lecture, Mendeleev (1889) not only attributed the success of the periodic law to Cannizzaro's ideas on the atomic theory (pp. 636–637), but went on to explicitly formulate the following hypothesis:

"the veil which conceals the true conception of mass, it nevertheless indicated that the explanation of that conception must be searched for in the masses of atoms; the more so, *as all masses are nothing but aggregations, or additions, of chemical atoms*" (Mendeleev, 1889, p. 640, emphasis added).

c. Again, at the Faraday lecture, Mendeleev (1889) took extreme care to explain the periodicity of properties of chemical elements on the basis of atomic theory: "periodic law has clearly shown that the masses of the atoms increase abruptly, by steps, which are clearly connected in some way with Dalton's law of multiple proportions . . . While connecting by new bonds the theory of the chemical elements with *Daltons's theory of multiple proportions, or atomic structure of bodies*, the periodic law opened for natural philosophy a new and wide field for speculation" (Mendeleev, 1889, p. 642, emphasis added). Interestingly, Mendeleev even seems to be considering the law of multiple proportions as synonymous with Dalton's atomic theory.

At this stage it is important to refer to Mendeleev's ambiguity/ambivalence towards the atomic theory. Throughout the nineteenth century, positivism was the dominant philosophy, which led all scientific work to be based strictly on experimental observations and all hypothetical propositions were considered speculative and hence non-scientific (Brush, 1976; Gavroglu, 2000; Holton, 1992). Mendeleev was clearly aware of this and on many occasions went out of his way to emphasize that the periodic "law itself was a legitimate induction from the verified facts" (Mendeleev, 1889, p. 639). Apparently, Mendeleev's dilemma was that on the one hand he could rightly claim that the periodic law was based on experimental properties of the elements (an aspiration of scientists in the late nineteenth century), and yet he could not give up the bigger challenge, namely, the possible causes of periodicity, and hence the importance of atomic theory.

Role of Predictions in the Development of the Periodic Table

Lakatos's (1970) idea of novel facts based on predictions is particularly helpful in understanding the periodic table. This work has been synthesized by Murphy (1989) and the role of novel facts can be conceptualized in the following terms: "A fact that played no role in the formulation of the theory is one whose existence, relevance to the theory, or interpretability in light of the theory was first documented after that theory was first proposed" (p. 388). Mendeleev's prediction of at least three elements (gallium, scandium, and germanium) and their respective properties, complies to a fair degree with this definition of novel facts (as suggested by Murphy). Interestingly, as early as 1877 (about two years after the discovery of gallium) a leading British chemist not only recognized Mendeleev's contribution as a theory but also referred to methodological issues: "The prevision of phenomena not yet observed has been rightly declared by methodologists to be one of the principal distinctions between science, in the strict sense of the term, and a mere accumulation of unorganized knowledge; the discovery of gallium thus shows the value of Mendeleev's theory" (Crookes, 1877, pp. 296, 298).

It is important to note that Mendeleev's contribution, in contrast to many of his predecessors, cannot be considered as a mere accumulation of knowledge, but rather has the basic elements of a scientific theory. Weisberg (2007) has argued cogently that Mendeleev had no empirical knowledge with respect to empty slots pertaining to missing elements. Consequently, he first needed to hypothesize the existence of the missing elements based on his theoretical framework and then make predictions.

Mendeleev is generally given credit for his successful predictions. Nevertheless, like any theoretical system, some of his predictions could not be confirmed, such as eka–niobiumand eka–caesium(cf. Brock, 1993). For some historians this could be the grounds for preferring accommodations rather than predictions. In the next section I discuss the relative importance of accommodations and predictions.

Relative Importance of Accommodations and Predictions as Possible Support for Mendeleev's Periodic Table

Availability of the atomic weights of about 60 elements enabled Mendeleev to accommodate the elements in the table according to various physicochemical properties (density, specific heat, atomic weight, atomic volume, melting point, valence, oxides, chlorides, and sulfides). In contrast to other discoverers, Mendeleev's work was characterized by the following aspects: the division into main and sub–groups, the vacant spaces left for undiscovered elements together with the prediction of some of their properties, i.e., the homologues of aluminum and silicon, the classification of the transition metals, and the reversal of tellurium–iodine (van Spronsen, 1969). Historians and philosophers of science continue to debate as to what was crucial for the acceptance of the periodic law by the scientific community: accommodation of the existing elements or the prediction of new ones (Brush, 1996; Dmitriev, 2004; Gordin, 2004; Kaji, 2003; Lipton, 1991, 2005; Maher, 1988). Lipton (1991) and Maher (1988) favor a predictivist thesis, namely, Mendeleev's law was accorded a greater recognition after the discovery of the first predicted element (gallium) in 1875. Mendeleev left various vacant spaces in his table and made many predictions and, of these, the following are the most important: (a) eka–aluminum(gallium); (b) eka–boron(scandium); (c) eka–silicon (germanium). Besides the atomic weights and physical properties (just presented), some of the chemical properties (formation of oxides, chlorides) of the predicted elements coincided to a remarkable degree with the discovered elements.

Mendeleev's Contribution: Theory or an Empirical Law?

There seems to be considerable controversy among philosophers of science with respect to the nature of Mendeleev's contribution: Was it a theory or an empirical law? Wartofsky (1968) clearly considers Mendeleev's contribution to be more than a simple empirical law:

> Mendeleev, for example, predicted that the blank space of atomic number 32, which lies between silicon and tin in the vertical column, would contain an element which was grayish–white,would be unaffected by acids and alkalis, and would give a white oxide when burned in air, and when he predicted also its atomic weight, atomic volume, density and boiling point, *he was using the periodic table as a hypothesis from which predictions could be deduced.* This was in 1871. *(p. 203, emphasis added)*

Actually, prediction of new elements by Mendeleev was no simple process. The atomic weight and the physicochemical properties had to adjust and follow the pattern dictated by elements preceding and following (in the period), and the elements above and below (in the group). These were stringent preconditions based on the hypothesis that the elements in the periodic table could be arranged in ascending order of their atomic weights provided they comply with the properties of neighboring elements in the periods/groups, and this precisely facilitated the prediction of new elements (novel facts in the Lakatosian framework).

Ziman (1978) recognizes the importance of predictions with respect to the persuasiveness (that is, heuristic power) of a theory, and hence Mendeleev's contribution can be considered as a theory:

> Needless to say, the most impressive way of validating a scientific theory is to confirm its *predictions* . . . the persuasive power of a successful prediction arises from the fact that it could not have been deliberately contrived. The most famous examples, such as Mendel'eef's prediction of the existence and properties of undiscovered elements to fill the gaps in the periodic table, or Gell–Mann and Neeman's prediction of the Omega–minusparticle to complete an SU(3) octet, have astonishing rhetorical power. *(p. 31, original emphasis)*

Shapere (1977) refers to the fact that historically Mendeleev's work has been referred to as a classification, system, table, or a law. Nevertheless, in his opinion the periodic table is neither a law nor a theory but, rather, an ordered domain:

> although the periodic table was widely referred to as a law, the general opinion of the time was that it could be called a "law" only in a rather loose sense, the true law being the precise mathematical expression of the "function" relating the atomic weights and the other properties of the elements. *(p. 536)*

Weisberg (2007) has clarified that if theories allow us to unify, make predictions, and frame explanations, then according to Shapere's own standard, Mendeleev must be considered as a theorist and his contribution as a theory.

This leads one to ask: Was Mendeleev an inductivist or a falsificationist? At this stage it is interesting to refer to difficulties faced by Mendeleev with respect to atomic weight reversals (e.g., nickel and cobalt; tellurium and iodine). Mendeleev in his periodic table had placed tellurium before iodine, knowing well that the atomic weight of tellurium was higher than that of iodine. Despite various efforts to determine the atomic weight of tellurium, it still remained higher than that of iodine. Based on the chemical properties (facts or theoretical framework if you prefer), Mendeleev insisted that tellurium must be placed before iodine. Similarly, the placing of an inert gas (argon) generated a lot of controversy, and constituted a crucial test for Mendeleev's theory. This clearly shows how placing of an element was not a straightforward question of ordering the elements in the ascending order of their atomic weights but, rather, Mendeleev used a hypothesis from which predictions could be deduced. In other words, his hypothesis was that tellurium must have certain physical and chemical properties, from which he could predict its place in the periodic table. As the atomic weight of tellurium then available did not allow it to be placed before iodine, Mendeleev, based on the strength of his hypothesis (theory), insisted that there must be an error in the determination of the atomic weight itself. Now, in order to respond to the question we raised at the beginning of this paragraph, let us seek help from Christie and Christie (2003, p. 173): "Had Mendeleev been an inductivist or a falsificationist, he would surely have had to abandon his scheme!" [authors are, of course, referring to Mendeleev having insisted that tellurium be placed before iodine, despite the empirical evidence of atomic weight to the contrary]. This clearly shows Mendeleev's *modus operandi*, based on the conviction of his theoretical framework he refused to accept the empirical evidence and insisted on the prediction that followed from his hypothesis and chemical properties of tellurium.

It appears that historians and philosophers of science generally conceptualize scientific progress to be dichotomous; experimental observations lead to scientific laws, which later facilitate the elaboration of explanatory theories. On the contrary, Lakatos (1970) has argued that

> the clash is not "between theories and facts" but between two high-level theories: between an *interpretative theory* to provide the facts and an *explanatory theory* to explain them; and the interpretative theory may be on quite as high a level as the explanatory theory. *(p. 129, original emphasis)*

In other words, scientific progress is characterized by a series of theories or models (plausible explanations), which vary in the degree to which they explain/interpret/predict the experimental findings. Consequently, we do not necessarily have to follow the theory/law (or for that matter ordered domain, codification scheme) dichotomy but, rather, it is reasonable to suggest that Mendeleev's work can be considered as an "interpretative" theory, which became "explanatory" (cf. Lakatos, 1970) after the periodic law was based on atomic numbers (based on Moseley,

1913, 1914; Thomson, Bohr, and Lewis). Furthermore, it is reasonable to suggest that an interpretative theory also provides explanation to a certain degree.

Based on the historical reconstruction of the periodic table it is plausible to elaborate the following aspects of Mendeleev's research program, namely, 'science in the making':

Mendeleev's theoretical framework (presuppositions): Mendeleev based his periodic law of chemical elements on Cannizzaro's ideas of atomic theory and that the veil that conceals the true conception of mass must be searched for in aggregations of chemical atoms. Furthermore, he considered Dalton's theory of multiple proportions (combination of elements), itself a manifestation of the atomic theory, and consequently used the atomic model as a framework for a new field of "speculation" and scientific research.

Research questions: (a) Can a simple arrangement of chemical elements according to their increasing atomic weight, facilitate a better understanding of the properties of substances? (b) Can this arrangement based on the periodic law help in the prediction and accommodation of new chemical elements?

Operationalizing heuristic principles: Accommodation of chemical elements into periods, groups, and sub–groupsbased on their atomic weights and various physico–chemical properties such as: density, specific heat, atomic volume, melting point, valence, oxides, chlorides, and sulfides. Vacant spaces were left for undiscovered elements together with the prediction of some of their properties. Furthermore, the atomic weight and the physicochemical properties had to adjust and follow the pattern dictated by elements preceding and following (in the period), and the elements above and below (in the group). These were stringent preconditions, namely, elements in the periodic table could be arranged in ascending order of their atomic weights provided they comply with the properties of neighboring elements in the periods and group.

Designing experiments: Based on his presuppositions, Mendeleev not only predicted new elements and their physicochemical properties but also suggested how these could be determined experimentally. For example, as the atomic weight of tellurium then available did not allow it to be placed before iodine, Mendeleev, based on the strength of his theoretical framework, insisted that there must an error, in the determination of the atomic weight itself. Similarly, corrections were also suggested in the atomic weights of beryllium, uranium, and other elements.

Understanding the nature of science: Of the 57 general chemistry textbooks analyzed (published in the USA, cf. Brito, Rodríguez, & Niaz, 2005) almost all recognized the importance of accommodation of elements in the periodic table. Again, 30 textbooks recognized the importance of prediction. Surprisingly, how–

ever, none of the textbooks recognized the relative importance of accommodation and prediction. Some textbooks came quite close to grasping Mendeleev's dilemma, and following is an example:

> More than 100 years later, we recognize Mendeleev's work as a milestone because he had the genius to realize that: a) there were many elements yet to be discovered and b) the characteristics of an element could be predicted from its atomic weight (and its position in his table). *(Kotz & Purcell, 1991, p. 334)*

This is a fairly good presentation of Mendeleev's dilemma. However, such presentations lack the appreciation that accommodation and prediction were alternative interpretations of the periodic table and hence controversial.

Most of the textbooks did not provide any explanation for periodicity, namely, by emphasizing: (a) inductive generalization; or (b) periodicity as a function of atomic theory. Following is an example of a textbook that came quite close to recognizing the role played by the atomic theory:

> Early in the nineteenth century, when Dalton's atomic theory was winning general acceptance, the first attempts were made toward classification of the elements into groups or families on the basis of similarities of physical and chemical properties . . . even in its primitive form as stated in 1869, this [periodic] law clearly pointed to regularities that hinted at an orderly subatomic structure of matter and provided a tremendous stimulus toward seeking to understand the internal structure of atoms, as chemists and physicists sought to construct an atomic model that would explain Mendeleev's generalization. *(Sisler, Dresdner, & Mooney, 1980, p. 150)*

This presentation has many novel features that can provide guidelines for future textbooks: (a) It establishes a relationship between Dalton's atomic theory and early attempts of classification of elements; (b) It points out the relationship between Mendeleev's first periodic table presented in 1869 and the subatomic structure of matter; and (c) development of the periodic table in itself provided a stimulus for the construction of an atomic model. Despite its merit, this presentation still lacks the understanding that inductive generalization based on physicochemical properties (accommodation) was considered by many scientists (also textbook authors and philosophers of science) to be an equally good explanation of periodicity in the table as was prediction.

Once again, almost all textbooks ignored the issue of whether Mendeleev's contribution was a theory or an empirical law. Following is an example of a textbook that considered Mendeleev's contribution to be an empirical law:

> Indeed, the periodic table is considered to be the single most useful study aid available for organizing information about the elements. For many

> years after the formulation of the periodic law and the periodic table, both were considered to be empirical. The law worked and the table was very useful, but there was no explanation available for the law or for why the periodic table had the shape it had. It is now known that the theoretical basis for both the periodic law and the periodic table lies in electronic theory. *(Stoker, 1990, p. 155)*

This presentation is quite representative of most textbooks. It ignores the fact that scientists were constantly trying to look for a "theoretical basis" of the periodic table, including Mendeleev himself.

Historical reconstruction of the periodic table provides a greater understanding of the following aspects of nature of science:

a. Alternative interpretations of a phenomenon are controversial and at the same time provide a greater understanding of scientific progress (based on the recognition of both accommodation and prediction in the development of the periodic table).
b. The explanation of periodicity in the periodic table as a function of an inductive generalization and/or atomic theory. This precisely highlights the role of atomic theory as a heuristic principle.
c. The tentative nature of scientific theories (based on the recognition that early attempts to understand the periodic table were also based on theoretical frameworks (theory of atomic weights) just as the modern atomic theory is based on atomic numbers).

Thomson's Determination of the Mass to Charge Ratio of Cathode Rays

It was the discovery of X-raysin 1895 that triggered Thomson and other physicists' interest in cathode rays. In the late nineteenth century when Thomson conducted his experiments he was well aware of the controversy with regards to the nature of cathode rays: Were they particles or were they waves in the ether? (Achinstein, 1991, pp. 299–300; Falconer, 1987, p. 243). Thomson (1897) summarized his experimental findings and the dilemmas involved in their interpretation in the following terms:

> The question next arises, What are these particles? Are they atoms, or molecules, or matter in a still finer state of subdivision? To throw some light on this point, I have made a series of measurements of the ratio of the mass of these particles to the charge carried by it. To determine this quantity, I have used two independent methods. *(p. 302)*

Later, Thomson (1897) reports the results of the mass to charge (m/e) ratio of the cathode ray particles:

> the value of m/e is independent of the nature of the gas, and its value 10^{-7} is very small compared with the value 10^{-4}, which is the smallest value of this quantity previously known, and which is the value for the hydrogen ion in electrolysis. *(p. 310)*

Thomson goes on to speculate that the smallness of m/e may be due to the smallness of m or the largeness of e, or to a combination of both (p. 310). A little later in the article Thomson suggests that the smallness of m/e is due to both (p. 312). In another important passage in the article, Thomson shares his thoughts with the reader:

> If, in the very intense electric field in the neighbourhood of the cathode, the molecules of the gas are dissociated and are split up, not into the ordinary chemical atoms, but into these primordial atoms, which we shall for brevity call corpuscles; and if these corpuscles are charged with electricity and projected from the cathode by the electric field, they would behave exactly like cathode rays. They would evidently give a value of m/e which is independent of the nature of the gas and its pressure, for the carriers are the same whatever the gas may be. *(p. 311)*

Thomson, of course, not only speculates and shares his thoughts with the reader but also suggests an explanation:

> The explanation which seems to me to account in the most simple and straightforward manner for the facts is founded on a view of the constitution of the chemical elements which has been favourably entertained by many chemists: *this view is that the atoms of the different chemical elements are different aggregations of atoms of the same kind*. In the form in which this hypothesis was enunciated by Prout, the atoms of the different elements were hydrogen atoms; in this precise form the hypothesis is not tenable, but if we substitute for hydrogen some unknown primordial substance X, there is nothing known which is inconsistent with this hypothesis. *(p. 311, emphasis added)*

As a grand finale, Thomson presents a theory of atomic structure:

> Thus on this view we have in the cathode rays matter in a new state, a state in which the subdivision of matter is carried very much further than in the ordinary gaseous state: a state in which all matter—that is, matter derived from different sources such as hydrogen, oxygen, etc.—is of one and the same kind; this being the substance from which all the chemical elements are built up. *(p. 312)*

Scrutinizing Thomson's 1897, article it can be observed that he goes far beyond a simple presentation of experimental results by speculating, hypothesizing, proposing models, offering explanations, and formulating a theory, which contrasts with the traditional view of the scientific method (cf. Niaz, 1994). This is what Schwab (1974) refers to when he says, "Most of us were taught a schoolbook version of a syntax under the guise of 'scientific method'" (p. 172). Thomson, although no philosopher of science, a few years later perhaps tried to respond to his critics in the following terms:

> From the point of view of the physicist, a theory of matter is a policy rather than a creed; its object is to connect or coordinate apparently diverse phenomena, and above all to suggest, stimulate and direct experiment. It ought to furnish a compass which, if followed, will lead the observer further and further into previously unexplored regions. *(Thomson, 1907, p. 1)*

Leon Cooper (Nobel laureate, physics, 1972) has summarized the reconstruction of Thomson's model of the atom in the following terms: "Not the first time that the greatest difficulty in doing a crucial experiment lay as much in developing necessary techniques as in its conception" (Cooper, 1970, p. 312; also Cooper, 1992, p. 312).

Based on a historical reconstruction it is plausible to elaborate the following aspects of Thomson's research program, namely 'science in the making':

Thomson's theoretical framework (presuppositions): Before starting his experiments, Thomson was fully aware of the controversial nature of cathode rays: Were they particles or waves in the ether? Based on a review of the literature he became convinced that cathode rays consisted of particles.

Research questions: After having decided with respect to the particulate nature of cathode rays (presupposition) Thomson formulated the following research questions: (a) What are these particles? (b) Are they atoms, or molecules, or matter in a still finer state of subdivision?

Operationalizing heuristic principle: Thomson clearly visualized that the determination of the mass (m) to charge (e) ratio (m/e) of the cathode rays would help him to identify the cathode ray particles as ions or as a universal charged particle.

Designing experiments: In his famous article, Thomson (1897) explicitly goes through the following stages of his research program: (a) Acceptance of the particulate nature of cathode rays (presupposition); (b) Formulation of research questions (What are these particles, atoms or molecules?); and (c) Elaboration of a heuristic principle, namely, constant mass to charge ratio (m/e) would provide evidence for the universal charged particle. Finally, Thomson provides details of his experiments and the results. Indeed, it seems that Thomson was writing for pedagogical purposes rather than for the scientific community. Actually,

Thomson was interested in pedagogical issues as he chaired the British Government's 1918 Report on *Natural Science in Education* (Thomson, 1918). This report emphasized the importance of history of science for science education (cf. Matthews, 1994a).

Understanding the nature of science: Cooper (1992) was the only general physics textbook (cf. Rodríguez & Niaz, 2004b) that presented a detailed presentation based on Thomson's (1897) original work, quite similar to the one outlined in Niaz (1998). Cooper's reconstruction follows the following sequence: Historical context → Thomson's understanding of the controversy → Experimental details → Thomson's strategy to tackle the dilemma → Mathematical details → Postulation of the electron. Similar presentations are also included in Cooper (1968, 1970).

One of the general chemistry textbooks (Mahan & Myers, 1987) also briefly mentioned the historical context with respect to the particle nature of cathode rays by providing a quote from Thomson (1897). Another textbook referred to Thomson's heuristic principle explicitly in the following terms:

> A very striking and important observation made by Thomson is that the *e/m* [charge to mass] ratio does not depend on the gas inside the tube or the metal used for the cathode or anode. The fact that the *e/m* ratio is the same whatever gas is present in the tube proves that the cathode ray does not consist of gaseous ions, for if it did, *e/m* would depend on the nature of the gas. *(Segal, 1989, p. 412)*

A Turkish general chemistry textbook (Yavuz, 1978) also presented the heuristic principle in similar terms (for details see Niaz & Coştu, 2009). A majority of the general chemistry and physics textbooks (Niaz, 2009a) ignored Thomson's heuristic principles and simply stated that based on his experiments he discovered the electron, which constitutes, a "rhetoric of conclusion" (Schwab, 1962, 1974).

Thomson's original work and some of the textbooks can facilitate a better understanding of 'science in the making' and eventually nature of science:

a. Based on their research questions, heuristic principles and experimental data, scientists often proceed by speculating, hypothesizing, proposing models, offering explanations, and formulating a theory, which contrasts with the traditional view of the scientific method.
b. A theory is more of a policy rather than a creed for suggesting, stimulating, and directing experiments for charting unexplored areas of study (based on Thomson, 1907, p. 1).
c. Difficulties involved in the design of an experiment lie as much in developing necessary techniques as in its conceptualization, namely heuristic principles (based on Cooper, 1992, p. 312).

Rutherford's Alpha Particle Experiments and the Nuclear Atom

It is generally believed that experiments of E. Rutherford and colleagues (Geiger & Marsden, 1909) led to the postulation of the nuclear atom, later published in the *Philosophical Magazine* (Rutherford, 1911). It is interesting to note that Rutherford had the experimental data as early as June 1909 (Geiger & Marsden, 1909), to postulate his model of the nuclear atom, and yet he did not do so until March 1911. Rutherford's decision to delay the publication of his atomic model reveals an important aspect of 'science in the making.' Interestingly, in his presidential address to the annual meeting of the British Association for the Advancement of Science (BAAS) held in August 1909 in Winnipeg (Canada), Rutherford was referring to the recent results of Geiger and Marsden (1909), when he reported:

> The conclusion is unavoidable that the atom is the seat of an intense electric field, for otherwise it would be impossible to change the direction of the particle in passing over such a minute distance as the diameter of a molecule. *(reproduced in Wilson, 1983, p. 292)*

A student may wonder why Rutherford presented the results from the experiments of Geiger and Marsden (1909) at the meeting and still refrained from presenting his model of the nuclear atom. Actually, it was not until March 1911 that Rutherford finally decided to present the nuclear model. This clearly shows that experimental data do not unequivocally lead to the formulation of scientific models and theories. Klassen and Dietrich (2010) have provided more background details of the 1909 BAAS meeting in Winnipeg, which can be helpful in motivating students' interest in science.

Soon after Geiger and Marsden (1909) published their results, Thomson and colleagues started working on the scattering of alpha particles in their own laboratory (published by Crowther, 1910). Although results from both laboratories were similar, the interpretations of Thomson and Rutherford were entirely different. Thomson propounded the hypothesis of *compound scattering*, according to which a large angle deflection of an alpha particle resulted from successive collisions between the alpha particle and the positive charges distributed throughout the atom. Rutherford, in contrast, propounded the hypothesis of *single scattering*, according to which a large angle deflection resulted from a single collision between the alpha particle and the massive positive charge in the nucleus. The rivalry between Thomson and Rutherford led to a bitter dispute between the proponents of the two hypotheses. At one stage, Rutherford even accused Crowther (1910), a colleague of Thomson, of having "fudged" the data in order to provide support for Thomson's model of the atom (Wilson, 1983). Heilbron (1981) has provided a succinct account of the rivalry between Thomson and Rutherford:

> Now in 1910 Thomson was the undisputed world master in the design of atoms, and neither Bragg nor Rutherford had yet tried their hand at the type of quantitative model making at which he excelled: they were challenging the acknowledged leader of English physics, their own former teacher, in what had been his most exclusive preserve. And Rutherford was coming into open conflict with Thomson's ideas for the first time. The situation thus did not lack a competitive aspect, and this, I think helps explain the harshness of the attacks against Crowther. *(Heilbron, 1981, p. 134).*

On the one hand, Rutherford was entirely convinced and optimistic that his model of the atom explained experimental findings better, and yet it seems that the prestige, authority, and even perhaps some reverence for his teacher made him waver. As compared to Thomson, Rutherford had almost no previous experience in the design of atomic models and still he could foresee and consequently interpret data in succinct terms:

> The observations, however, of Geiger and Marsden (1909) on the scattering of alpha rays indicate that some of the alpha particles must suffer a deflexion of more than a right angle at a single encounter. They found, for example, that a small fraction of the incident alpha particles, about 1 in 20,000 were turned through an average angle of 90° in passing through a layer of gold-foil. . . A simple calculation based on the theory of probability shows that the chance of an alpha particle being deflected through 90° is vanishingly small . . . A simple calculation shows that *the atom must be a seat of an intense electric field in order to produce such a large deflexion at a single encounter. (Rutherford, 1911, p. 669, emphasis added)*

These considerations helped Rutherford in the design of experiments and the postulation of his model.

Based on a historical reconstruction it is plausible to elaborate the following aspects of Rutherford's research program, namely 'science in the making':

Rutherford's theoretical framework (presuppositions): Before starting his research, Rutherford accepted Thomson's model of the atom (plum-pudding)as his theoretical framework, considered to be the dominant paradigm until almost 1910. However, after receiving results from his colleagues (Geiger & Marsden, 1909), the nuclear model constituted his theoretical framework. This provides an example of how scientists can accept changes in their presuppositions.

Research questions: (a) Can alpha particles be deflected through large angles (Rutherford frankly admitted that he did not expect this to happen)?; and (b) What are the consequences of large angle deflections for atomic models?

Operationalizing heuristic principles: Probability of large angle deflection of an alpha particle was extremely small, hence the need for a massive positive charge in the atom and consequently the need for a change in the existing atomic models.

Designing experiments: Rutherford faced considerable difficulty while designing and interpreting results from the experiments. First he had results from his own laboratory (Geiger & Marsden, 1909) and later from a colleague of Thomson (Crowther, 1910). Rutherford was convinced that results from the latter were "fudged" in order to provide support for Thomson's model of the atom.

Understanding the nature of science: Once again Cooper (1992) introduces the Rutherford model in a historical context:

> The situation in 1910, as pieced together from these various *experiments* and *conjectures*, seemed to be somewhat as follows: atoms, electrically neutral, consisted of electrons (of very small mass) and positive material with the major part of the mass. *(p. 316, italics added)*

Cooper's conjugation of *experiments* and *conjectures* is indeed a novel way of conceptualizing scientific development. In other words, experimental data in themselves do not dictate theory, but rather it is the *heuristic principle* (underlying idea) coupled with experimental data that helps a scientist to conceptualize future developments. Later, Cooper goes on to present the crucial argument in support of the Rutherford model of the atom:

> Rutherford calculated that from the large Thomson positive charge distribution particles should never be deflected more than 0.03 degrees in a single collision; in undergoing multiple collisions they should have about an equal chance of being deflected one way as another. Therefore, large deflections as a result of many single deflections in the same direction were very improbable. (It had been calculated on the basis of the Thomson model that a total deflection greater than 90° in traversing the gold foil would have only one chance in 10^{3500} of occurring.) *(p. 321)*

Another general physics textbook also presented Rutherford's heuristic principle based on single scattering in a fairly lucid and educationally helpful manner.

> The Thomson model for scattering fails when we examine the probability for scattering at large angles. If each individual scattering deflects the projectile through an angle of around 0.01°, then to observe projectiles scattered through a total angle greater than 90°, we must have about 10^4 successive scatterings, *all* of which push the projectile toward larger angles. Since the

> probabilities of individual scatterings toward either larger or smaller angles are equal, the probability of having 10^4 successive scatterings toward larger angles, like the probability of finding 10^4 successive heads in tossing a coin, is about $(1/2)^{10.000} = 10^{-3000}$. *(Krane, 1996, p. 178, original italics)*

The example of tossing a coin in order to understand the probability argument can be very helpful for students. This makes the understanding of Thomson's refusal to accept Rutherford's model of the atom even more difficult. At this stage it would be interesting to consider the following questions:

1. Thomson was well–versedin both mathematical physics and atomic modeling and still he refused to accept Rutherford's arguments based on probability the–ory. In contrast Rutherford had to enroll in early 1909 to attend elementary lectures on probability given by Horace Lamb, and his notebooks bear witness to his attendance and having taken extensive notes (Wilson, 1983, p. 290).
2. Both Thomson and Rutherford were well known to each other (both as teacher–student and later colleagues) and could easily have met over dinner and reconciled their differences. This clearly shows that scientists are reluc–tant to give up their heuristic principles (cf. hard–coreof beliefs, Lakatos, 1970; guiding assumptions, Laudan, Laudan, & Donovan, 1988).

These arguments played a crucial role in the bitter Thomson–Rutherford dis–pute (Heilbron, 1981; Wilson, 1983) and are generally ignored by most text–books. Rutherford's experiments on alpha particles, postulation of the nuclear atom and the ensuing controversy with Thomson can enhance our understanding of the NOS, by taking the following into consideration: (a) Scientists generally are reluctant to give up and change their heuristic principles (based on hard–core of beliefs); (b) Postulation and acceptance of a new model is difficult due to the rivalry with the existing models; and (c) Conjugation of *experiments* and *conjectures*, namely the heuristic principles, help scientists to understand and conceptualize experimental data (based on Cooper, 1992, p. 316).

Bohr's Model of the Atom

In his now famous trilogy Bohr (1913) starts on a controversial note by point–ing out the difficulties associated with the stability of Rutherford's model of the atom:

> In an attempt to explain some of the properties of matter on the basis of this atom–model[Rutherford's] we meet, however, with difficulties of a seri–ous nature arising from the apparent instability of the system of electrons: difficulties purposely avoided in atom–modelspreviously considered, for instance, in the one proposed by Sir J.J. Thomson. *(p. 2)*

Next, Bohr formulates his famous postulate of *quantum of action*. Contrary to textbook accounts, Bohr's major objective was not to explain the Balmer and Paschen series of the hydrogen line spectrum but, rather, to explain the paradoxical stability of the Rutherford atom. According to Lakatos (1970): "Bohr's problem was not to explain Balmer's and Paschen's series, but to explain the paradoxical stability of the Rutherford atom. Moreover, Bohr had not even heard of these formulae before he wrote the first version of his paper" (p. 147).

The "first version" Lakatos is referring to is of course the "Rutherford Memorandum" written by Bohr for Rutherford in June–July of 1912, and is considered by Heilbron and Kuhn (1969) to be a crucial document. A letter, written by Bohr to Rutherford on January 31, 1913, shows that even then he was not fully aware of the implications of spectroscopic research for his problem [reproduced by Rosenfeld, 1963, pp. xxxvi–xxxvii]. A reconstruction of the events that led Bohr to postulate his model of the atom has important implications for our understanding of scientific progress and practice. According to Lakatos (1970):

> Since the Balmer and the Paschen series were known before 1913 [year of Bohr's first publication], some historians present the story as an example of a Baconian "inductive ascent": (1) the chaos of spectrum lines, (2) an "empirical law" (Balmer), (3) the theoretical explanation (Bohr). *(p. 147)*

The inductive ascent refers to the positivist/inductivist interpretation of the history of science, according to which scientific progress follows the sequence, experimental observations, scientific laws, and theories. Another important aspect of Bohr's model of the atom is the presence of a deep philosophical chasm: that is, in the stationary states, the atom obeys classical laws of Newtonian mechanics; on the other hand, when the atom emits radiation, it exhibits discontinuous (quantum) behavior. On the basis of these and other arguments, Bohr's (1913) article, in general, had a fairly adverse reception in the scientific community.

Based on a historical reconstruction it is plausible to elaborate the following aspects of Bohr's research program, namely 'science in the making':

Bohr's theoretical framework (presuppositions): (1) The electron in an atom has only certain definite stationary states (fixed energy) of motion allowed to it; (2) When an atom is in one of these states it does not radiate energy. When an electron in an atom changes from a high-energystate to a state of lower-energy,the atom emits a quantum of radiation whose energy ($h\upsilon$) is equal to the difference in the energy of the two states; (3) In any of these stationary states the electron moves in a circular orbit about the nucleus; and (4) The states of allowed electronic motion are those in which the angular momentum of the electron is an integral multiple of $h/2\pi$.

Research question: How can the paradoxical stability of the Rutherford model of the atom be explained?

Operationalizing heuristic principles: In Bohr's case there was a close relationship between his heuristic principles and his four postulates of which the "quantum of action" was the most controversial. At first Bohr was not aware of the implications of atomic spectra for his research program. Once his attention was drawn by a spectroscopist (H.M. Hansen) he became quite enthusiastic about experimental support for his model based on atomic spectra (cf. Jammer, 1966, p. 77).

Designing experiments: Bohr's model, however, was based on too simple an assumption of circular orbits all in a plane. In the Bohr–Sommerfeld model of the atom (Sommerfeld, 1916), electrons moved in Keplerian ellipses and during their orbits, they penetrated the region of internal electrons, thereby causing a coupling of the revolving electrons. To Bohr's satisfaction, Paschen's (1916) experimental measurements of the helium spectrum were in agreement with the Bohr–Sommerfeld model (cf. Niaz & Cardellini, 2011).

Understanding the nature of science: Cooper (1992) included the heuristic principle in terms that can help to elucidate the historical aspects in general physics textbooks:

> In 1913 Niels Bohr proposed his famous theory of the hydrogen atom. One cannot say that he resolved the problems raised by Rutherford. In a sense he crystallized the dilemma in an even more dramatic form. Focusing his attention entirely on the construction of a nuclear atom, Bohr took what principles of classical physics he needed and added several nonclassical hypotheses almost without precedent; the mélange was not consistent. But they formed a remarkably successful theory of the hydrogen atom. It would be years before it could be said that one had a consistent theory again. *(p. 325)*

It would be interesting to highlight the following aspects of this presentation: (a) Although Bohr's theory is famous, it still did not resolve the paradoxical stability of the Rutherford atom but, rather, formulated a new dilemma; (b) Bohr mixed various principles of classical and quantum physics to present a successful theory of the hydrogen atom; and (c) Scientific theories can be successful (e.g., Bohr's) and still be based on an inconsistent foundation.

A general chemistry textbook presented an interesting perspective:

> There are two ways of proposing a new theory in science, and Bohr's work illustrates the less obvious one. One way is to amass such an amount of data

> that the new theory becomes obvious and self–evidentto any observer. The theory then is almost a summary of the data. The other way is to make a bold new assertion that initially does not seem to follow from the data, and then to demonstrate that the consequences of this assertion, when worked out, explain many observations. With this method, a theorist says, "You may not see why, yet, but please suspend judgment on my hypothesis until I show you what I can do with it." Bohr's theory is of this type. *(Dickerson, Gray, Darensbourg, & Darensbourg, 1984, p. 264)*

An important aspect of this textbook's presentation is that it provides students with an insight with respect to alternative approaches to theory construction. Furthermore, it goes beyond the traditional empiricist epistemology in which theories are dictated by empirical data. On the contrary, Dickerson et al. (1984) use Bohr's methodology as an example of how sometimes scientists do not have all the necessary data and consequently they do not discard a promising theory but, rather, suspend judgment (cf. Holton, 1978, suspension of disbelief, in the context of the oil drop experiment).

Historical reconstruction of Bohr's model of the atom, as an example of 'science in the making' provides important guidelines for implementing the NOS: (a) Scientific theories at times do not solve all the problems but, rather, create new dilemmas; (b) Scientific theories can be successful and still based on an inconsistent foundation; (c) In the early periods of theory construction, a scientist needs to hold in abeyance final judgments, lest a promising theory be discarded (based on Holton, 1978, p. 212). Furthermore, Holton (1978) has recommended that every novice be taught that, "the graveyard of science is littered with those who did not suspend disbelief while the data were pouring in" (p. 212).

Millikan's Determination of the Elementary Electrical Charge

Robert Andrews Millikan was perhaps the most successful experimental physicist of the twentieth century. Besides J.J. Thomson's (1897) work on the nature of cathode rays, Millikan was deeply impressed by the potential offered by the study of the properties of gases ionized by X–rays(as compared to ionization in electrolysis):

> But the ionization produced in gases by X–rayswas of a wholly different sort, for it was observable in pure gases like nitrogen or oxygen, or even in monoatomic gases like argon and helium. Plainly, then the neutral atom even of a monoatomic substance must possess minute electrical charges as constituents. *(Millikan, 1917, p. 40)*

This provided Millikan with important insight with respect to the complex structure of the atom based on electrical charges. Early in his career he summarized the development of atomic structure at the turn of century, in the following terms (Millikan, 1917, p. 41):

1. What are the masses of the constituents of the atoms torn asunder by X–raysand similar agencies?
2. What are the values of the charges carried by these constituents?
3. How many of these constituents are there?
4. How large are they, i.e., what volumes do they occupy?
5. What are their relations to the emission and absorption of light and heat waves, i.e., of electromagnetic radiation?
6. Do all atoms possess similar constituents? In other words, is there a primordial sub–atomout of which atoms are made?

This could very well be considered Millikan's research program and interestingly the first question refers to the potential offered by X–rays.The sixth question, of course referred to Thomson's finding that the charge to mass (*e*/*m*) ratio of cath–ode rays is independent of the gas in the discharge tube, and recognized as such by Millikan (1917): "This was an indication of an affirmative answer to the sixth question above" (p. 42). It precisely set the stage for Millikan's determination of the elementary electrical charge (I have presented the experimental details elsewhere, cf. Niaz, 2000b, 2005a, 2009a). This enabled Millikan (1917) to elaborate the atomic nature of electricity as a heuristic principle of his research program:

> whether electricity in gases and solutions is actually built up out of electrical atoms, each of which has exactly the same value, or whether the electron which had first made its appearance in Faraday's experiments on solutions and then in Townsend's and Thomson's experiments on gases is after all only a *statistical mean* of charges which are themselves greatly diver–gent. *(p. 58, original italics)*

According to Holton (1978) it is important to differentiate between the design and the discovery of the oil drop experiment. Millikan evidently did not design the experiment but, rather, discovered it. It was the electron theory which sug–gested the existence of the elementary electrical charge (*Millikan's heuristic principle*) and hence the need for its experimental determination. Nagel (1961) has explained this difference in cogent terms:

> It is unlikely that Millikan (or anyone else) would have devised [discovered] the oil–dropexperiment if some atomistic theory of electricity had not first suggested a question that seemed important in the light of the theory and that the experiment was intended to settle. *(p. 90)*

After many years of hard work and a running controversy with Felix Ehrenhaft (University of Vienna), Millikan was able to convince the scientific community that the charge on the electron is not a *statistical mean* but, rather, a universal charged particle with a definite and discrete "granular" structure.

Based on a historical reconstruction it is plausible to elaborate the following aspects of Millikan's research program for determining the elementary electrical charge, namely, 'science in the making':

Millikan's theoretical framework (presuppositions): All atoms have a universal charged particle based on Thomson's determination of the charge to mass ratio (found to be constant) of cathode rays emitted by different metals.

Research questions: (a) Do all atoms possess similar constituents? and (b) Is there a primordial sub–atomout of which atoms are made? (cf. Millikan, 1917, p. 41). Of course, a positive response would mean that all atoms have the universal charged particle (electron).

Operationalizing heuristic principles: The charge on the oil drop could have been a statistical mean of particles having varying charges or a discrete particle with a definite charge in all experiments.

Designing experiments: Millikan did not design the experiment but, rather, discovered it. It took many years to refine the experimental procedure of the oil drop experiment and then a running controversy with Felix Ehrenhaft made the interpretation of the data extremely difficult (for details see Holton, 1978; Niaz, 2005a).

Understanding the nature of science: Klassen (2009) has shown that the oil drop experiment needs to be performed in the laboratory within an HPS perspective. Heering and Klassen (2010) have developed a new collaborative approach to facilitate students' understanding of the oil drop experiment. These authors have suggested a shift in focus, namely, that students do not have to determine the elementary electrical charge (that was Millikan's problem) but, rather, demonstrate the elementary charge, which is a more plausible educational objective. Based on this shift in perspective students were found to have a better understanding of 'science in the making.'

Of the 43 general physics textbooks analyzed (published in the USA) within an HPS perspective (Rodríguez & Niaz, 2004c) only two (Olenick, Apostol, & Goodstein,1985; Urone, 2001) mentioned Millikan's *heuristic principle* in the context of his guiding assumption. One of the physics textbooks went to considerable length (about five pages) to present Millikan's research methodology and pointed out the dilemmas and contradictions in the handling of the data: "By observing the motion of the hundreds of droplets with different charges on them,

Millikan *uncovered the pattern he expected*: the charges were multiples of the smallest charge he measured" (Olenick et al., 1985, p. 241, italics added). "The pattern" can be considered as an oblique reference to Millikan's guiding assumption. The textbook reproduced the following quote from Millikan's laboratory notebook (dated 15 March, 1912; see Holton, 1978 for Millikan's lab. notebooks): "One of the best ever [data] . . . almost exactly *right*. Beauty—publish" (original italics). After reproducing the quote, the textbook authors asked a very thought provoking question: "What's going on here? How can it be right if he's supposed to be measuring something he doesn't *know*? One might expect him to publish everything!" (Olenick et al., 1985, p. 244, original italics). These are important issues related to the NOS, namely, can a scientist know beforehand what he is going to find and, what is even more difficult to understand, how can a scientist know the right answer before doing the experiment? Interestingly, the authors themselves provided further insight and advice for students:

> Now, you shouldn't conclude that Robert Millikan was a bad scientist . . . What we see instead is something about how real science [cutting–edge]is done in the real world. What Millikan was doing was not cheating. He was applying scientific judgment . . . But experiments must be done in that way. Without that kind of judgment, the journals would be full of mistakes, and we'd never get anywhere. So, then, what protects us from being misled by somebody whose "judgment" leads to wrong results? Mainly, it's the fact that someone else with a different prejudice can make another measurement . . . Dispassionate, unbiased observation is supposed to be the hallmark of the scientific method. Don't believe everything you read. Science is a difficult and subtle business, and there is no method that assures success. *(Olenick et al., 1985, p. 244)*

This is a good illustration of how Millikan's presuppositions and heuristic principle can facilitate students' understanding of 'science in the making' and, eventually, the NOS: (a) Doing experiments means gathering data and its interpretation (scientific judgment); (b) Without such judgments journals would be full of mistakes; (c) Some scientists can be misled in their judgments; (d) Another scientist with a different heuristic principle can present an alternative interpretation, that is, science is self–correcting;(e) There are no dispassionate, unbiased observations as suggested by the scientific method: observations are theory laden.

Millikan's Determination of Planck's Constant *h* (Photoelectric Effect)

Starting in 1902, Lenard conducted important experiments on the photoelectric effect, and explained that electrons in an atom already have the necessary potential energy and incident light only triggers the release of the selected electrons.

According to Wheaton (1983): "Until 1911 this *triggering hypothesis* formed the basis of almost all physicists' understanding of the photoelectric effect" (p. 75, original italics). A major criticism of the triggering hypothesis was due to the absence of an influence of temperature on the photoelectric effect. In order to explain the photoelectric effect, Einstein (1905) proposed that ordinary light behaves as though it consists of a stream of independent localized units of energy that he called *lightquanta*. Thus, if light consists of localized quanta of energy, an electron in an atom will receive energy from only one lightquantum at a time. In 1913, Planck attempted to interpret the photoelectric effect by suggesting that light energy is not transformed into electron kinetic energy as the latter already exists within the atom. Again, according to Wheaton (1983): "Planck's theory was the quantum analog of the triggering hypothesis" (p. 179) and thus suffered most of its drawbacks. It is important to note that in the first decades of the twentieth century, Lenard's trigger hypothesis, Planck's quantum analog of the trigger hypothesis, and Einstein's quantum hypothesis constituted rival hypotheses to explain the same photoelectric phenomenon. However, Einstein's hypothesis explained more experimental facts and thus provided greater heuristic power (Lakatos, 1970).

Besides the determination of the elementary electrical charge, Millikan devoted considerable effort to the experimental determination of Einstein's photoelectric equation (see Equation 1 below) and, as a consequence, calculated the experimental value of Planck's constant h (Millikan, 1914, 1915, 1916a, 1916b). Millikan's value for h (6.57×10^{-27} erg. sec.) came quite close to that reported by Planck (6.55×10^{-27}).

$$\tfrac{1}{2}mv^2 = Pe = hv - p \qquad \text{Equation 1}$$

(Where p = energy used to release the electron from the metal, e = charge of the electron, P = stopping potential, v = light frequency, $\tfrac{1}{2}mv^2$ = maximum kinetic energy of the ejected electrons.)

Millikan was quite satisfied with all the technical innovations that he developed for his experimental measurements and concluded "so far as experiment has now gone, Einstein's photoelectric equation, whatever may be said of its origin, seems to stand up accurately under all of the tests to which it has been subjected" (Millikan, 1916a, p. 31). Later critical evaluation of Millikan's experimental determinations also recognized its importance: "Millikan's results were never questioned, and were quickly recognized by leading European physicists to be definitive. But acceptance of the Einstein law of the photoelectric effect did not carry with it acceptance of the hypothetical lightquantum" (Wheaton, 1983, p. 241). This aspect of Millikan's interpretation of Einstein hypothesis of lightquanta is generally ignored by science textbooks.

It is interesting to note that in the same publication, Millikan (1916b) recognized the validity of the Einstein photoelectric equation and simultaneously questioned the underlying hypothesis of lightquanta: "Despite then the apparently complete success of the Einstein equation, the physical theory of which it was designed to be the symbolic expression is found so untenable that Einstein himself, I believe, no longer holds to it. But how else can the equation be obtained?" (Millikan, 1916b, p. 384). This may sound incredulous to any student of the HPS. Actually, Millikan went far beyond doubting Einstein's own development since 1905 (Einstein having restated his thesis in 1916), and it is worthwhile to follow Millikan's train of thought in some detail:

> It was in 1905 that Einstein made the first coupling of photo effects . . . with any form of quantum theory by bringing forward the bold, not to say *the reckless, hypothesis* of an electro–magneticlight corpuscle of energy $h\nu$, which was transferred upon absorption to an electron. This hypothesis may well be called reckless first because an electromagnetic disturbance which remains localized in space seems a violation of the very conception of an electromagnetic disturbance, and second because it flies in the face of the *thoroughly established facts of interference*. The hypothesis was apparently made solely because it furnished a ready explanation of one of the most remarkable facts brought to light by recent investigations, viz., that the energy with which an electron is thrown out of a metal by ultra–violetlight or X–raysis independent of the intensity of the light while it depends on its frequency. This fact alone seems to demand *some modification of classical theory* or, at any rate, it has *not yet been interpreted satisfactorily in terms of classical theory. (Millikan, 1916b, p. 355, italics added)*

It is not farfetched to suggest that Millikan's train of thought could be understood along the following lines: We have thoroughly established facts of interference based on the classical wave theory → although, some modification of classical theory may be necessary → nevertheless, Einstein's hypothesis of lightquanta is reckless. This clearly shows that Millikan's presuppositions were guided by the classical wave theory of light and despite experimental evidence to the contrary he resisted changes in the heuristic principle of his research program. Millikan's heuristic principle at this stage could be understood in the following terms: Classical wave theory can explain established facts of interference, and it is expected that it will eventually (with some modifications) be able to explain the photoelectric effect.

Millikan was awarded the 1923 Nobel Prize in physics for his contributions to the experimental determination of both the elementary electrical charge and Planck's constant based on the Einstein photoelectric equation. In his acceptance speech delivered on May 23, 1924, Millikan recounted the difficulties during the experiments and expressed reservations with respect to the hypothesis of lightquanta.

In retrospect, Holton (1999) has raised a very pertinent point: "So Millikan's [1916b] paper is not at all, as we might now naturally consider it to be, an experi–

mental proof of the quantum theory of light" (p. 232). This statement has important implications. The 1916 publication was considered by its own author as an experimental test of Einstein's photoelectric equation and in no way a confirmation of the underlying hypothesis of lightquanta (Millikan, 1916b). On the contrary, most textbooks and physicists at present would consider Millikan's experimental data as evidence for Einstein's quantum theory (hypothesis of lightquanta). How is one to resolve this dilemma? A plausible explanation is provided by Millikan's commitment to his presuppositions (also heuristic principle) with respect to the wave theory of light stated quite explicitly in various publications (Millikan, 1916a, 1916b, 1965). Furthermore, various scholars have corroborated Millikan's adherence to the wave theory of light (Holton, 1999; Wheaton, 1983). Stuewer (1975) has explained that Millikan was not alone in his rejection of Einstein's theory (milieu of the time): "Therefore, in common with Lorentz, Thomson, Sommerfeld, and Richardson, Millikan did not take the validity of Einstein's *equation*—which he established beyond doubt—to constitute proof of the validity of Einstein's light quantum *hypothesis*" (p. 75, original italics).

Holton (1999) refers to the fact that Millikan published his *Autobiography* in 1950 at the age of 82. It is interesting what Millikan had to say about experimental data with respect to Einstein's photoelectric equation, which:

> proved simply and irrefutably, I thought, that the emitted electron that escapes with the energy hν gets *that energy by the direct transfer of hν units of energy from the light to the electron* and hence scarcely permits of any other interpretation than that which Einstein had originally suggested, namely that of the semi-corpuscular or photon theory of light itself. *(Millikan, 1950, pp. 101–102, emphasis in the original)*

After this, Holton (1999) refers to this as the first irony in the following terms: "In the end, Millikan re-imaginedthe complex personal history of his splendid experiment to fit the simple story told in so many of our physics textbooks" (p. 236). Interestingly, Stuewer (1975, p. 88) considers this adjustment on the part of Millikan as "shocking," considering the fact that even in 1924, in his Nobel Prize acceptance speech, Millikan still questioned Einstein's hypothesis of lightquanta.

In retrospect it is interesting to observe that even Millikan with an excellent background and training in the determination of both *e* (elementary electrical charge; Millikan, 1913) and *h* (Planck's constant; Millikan, 1916b), could not fathom beyond the "observable empirical data," and give up his prior presuppositions (Holton, 1999; Wheaton, 1983). One would have expected Millikan to at least consider the quantum theory on the same footing as wave theory, from, say, 1916 onwards. This shows that despite the importance of empirical data, presuppositions play a crucial role in a scientist's struggle in order, at times, to understand experimental data (Planck's constant, *h*) and on occasions even to discard data (*e*, elementary electrical charge).

According to Lakatos (1970) a scientist generally does not abandon his "hard–core" in the face of anomalous data. This is exactly what Millikan did in the case of the oil drop experiment and it paid good dividends (cf. Holton, 1978). In his photoelectric experiments, Millikan followed the same strategy that is a strong belief in the classical wave theory of light (presupposition). In this case, however, Millikan's perseverance with the presupposition did not bear fruit and he ultimately had to change the heuristic principle of his research program. It is, however, somewhat disconcerting that Millikan had nothing to say with respect to his earlier criticism of Einstein's hypothesis of lightquanta.

The relationship between Millikan's two major research programs (oil drop experiment and photoelectric effect) and the role played by presuppositions facilitate our understanding of 'science in the making,' and that in some cases a scientist can even change their heuristic principles. This shows clearly that presuppositions (right or wrong) form an integral part of the research program of a scientist and that at times it requires a considerable amount of intellectual effort to provide grounds for the "hard–core"to crumble and make way for the new framework. Both Duhem (1914) and Lakatos (1970) have emphasized that the "hard–core" (based on presuppositions) is not only difficult to refute but it also develops slowly and not like Athene from the head of Zeus. Millikan believed in the wave theory of light without any reservations (despite experimental evidence to the contrary) and was skeptical of the lightquanta (quantum hypothesis) as late as 1924. Holton's (1999) ironies provide an interesting historical backdrop to Millikan's attempt in his autobiography to "rewrite" history to make his account more in tune with the unfolding of historical events after 1924.

Based on a historical reconstruction it is plausible to elaborate the following aspects of Millikan's determination of Planck's constant (photoelectric effect), namely, 'science in the making':

Millikan's theoretical framework (presupposition): Millikan strongly believed in the classical wave theory of light, which could explain the well–knownphenomenon of interference and hence apparently could also explain the photoelectric effect.

Research question: Can Einstein's photoelectric equation (despite the questionable nature of the theory) facilitate calculation of Planck's constant, h?

Operationalizing heuristic principle: Classical wave theory can explain established facts of interference, and it is expected that it will eventually (with some modifications) be able to explain the photoelectric effect. Eventually Millikan had to give up the classical theory and accept Einstein's lightquanta, which shows how scientists can also change their heuristic principles.

Designing experiments: After working on the photoelectric effect for many years, Millikan (1916b) was able to design the experiment and calculate Planck's constant, h, which was soon accepted by the scientific community.

Understanding the nature of science: Of the 103 general physics textbooks analyzed (published in the USA, between the 1950s and the 2000s) only five made a brief mention of Millikan's presupposition and belief in the classical wave theory of light (cf. Niaz, Klassen, McMillan, & Metz, 2010a). Very few textbooks mention that Millikan's experimental data provided support for Einstein equation but not his theory and the following is an example, "[n]ot until 1914–1915 would Robert Millikan finally and conclusively determine that the Photoelectric Equation was in complete agreement with experiment" (Hecht, 1998, p. 1028). Again, none of the textbooks mentioned that scientific theories are underdetermined by experimental evidence. One of the textbooks came quite close to referring to the dilemma faced with respect to the lack of acceptance of Einstein's quantum hypothesis in the scientific community:

> Although Millikan showed that this equation agreed with his experiment in every detail he himself remained unconvinced that Einstein's light particles were real. He wrote of Einstein's ". . . bold, not to say reckless, hypothesis . . ." and wrote further that Einstein's photon concept ". . . seems at present to be wholly untenable." Planck, the very originator of the constant, *h*, did not at once accept Einstein's photons either. In recommending Einstein for membership in the Royal Prussian Academy of Sciences in 1913 he wrote: "that he may sometimes have missed the target in his speculations, as for example in his theory of light quanta, cannot really be held against him." Far from being unusual it is almost commonplace that truly novel ideas are accepted only slowly, even by such men of genius as Millikan and Planck. *(Halliday & Resnick, 1981, p. 783)*

Of the 118 general chemistry textbooks analyzed (published in the USA between the 1950s and the 2000s) none referred to Millikan's presupposition and belief in the classical wave theory of light (Ospina, 2010). Only two textbooks mentioned that Millikan's experimental data provided support for Einstein's equation and not his theory, and the following is an example: "Robert Millikan, the experimenter who designed the ingenious oil drop experiment for determining the charge of the electron, verified Einstein's photoelectric equation in 1916 after careful measurements" (Fine & Beall, 1990, p. 478). Very few textbooks referred to the problematic nature of Einstein's hypothesis of lightquanta. The following is an example:

> Einstein's theory with respect to light constituted a dilemma for the scientists. On the one hand the theory explained the photoelectric effect satisfactorily. However, the particulate nature of light was not consistent with known wave properties. The only way of resolving this dilemma was to accept the idea that light possessed the properties of both particle and wave. Depending on the type of experiment, light behaved as a wave or a stream

of particles. This concept differed radically from what the physicists thought were matter and radiation, and it took a long time to be accepted. *(Chang, 2007, p. 274)*

Presentation of this dilemma with respect to the explanation of the photoelectric effect can arouse students' curiosity and understanding with respect to the difficulties involved in the development of scientific theories. Based on these examples from general physics and chemistry textbooks it is plausible to suggest that the inclusion of the following aspects related to the photoelectric effect can facilitate a better understanding of the dynamics of 'science in the making': (a) Einstein's hypothesis of lightquanta constituted a rival to Lenard's trigger hypothesis; (b) Einstein's hypothesis was not accepted by the scientific community, including Planck, the "originator" of the quantum hypothesis, for many years; (c) Millikan presented experimental evidence to support Einstein's photoelectric equation and still rejected his quantum hypothesis until about 1924; and (d) Millikan's prior belief in the classical wave theory of light was the major reason for his rejection of Einstein's hypothesis of lightquanta.

Both Einstein and Millikan's heuristic principles can facilitate a better understanding of the NOS: (a) Millikan interpreted data from photoelectric phenomenon based on the classical wave theory of light, whereas Einstein interpreted the same data based on the quantum hypothesis. This shows the theory-laden nature of observations and that the same data can be interpreted within different conceptual frameworks; (b) Rival hypotheses were put forward to explain the same experimental data (Lenard's trigger hypothesis, Planck's quantum analog, and Einstein's quantum hypothesis); (c) Underdetermination of scientific theories by experimental evidence (Millikan's experimental data did not necessarily prove Einstein' theory but rather provided supported for his equation); and (d) The tentative nature of scientific knowledge (varying interpretations of the photoelectric effect based on different theoretical frameworks); (e) Despite resistance, heuristic principles of a scientist can change over the years as a consequence of a constant process of critical appraisals (Millikan, despite initial belief in the wave theory of light, finally did accept Einstein's quantum hypothesis).

Determination of Wave-Particle Duality by de Broglie

Postulation and understanding of wave-particleduality was a controversial topic from the very beginning and is closely enmeshed with the origin and development of the photoelectric effect and quantum theory. The objective of this section is to present a historical reconstruction of the wave-particleduality and the roles played by de Broglie, Einstein, and Schrödinger, and the experimental evidence provided by Davisson and Germer (1927).

Working on the assumption that the classical theory of radiation needed to be radically revised, Einstein (1909a, 1909b) published two important papers in

1909, in which he argued explicitly for a fusion of the wave and emission theories of light. This support for the wave–particleduality, however, was not accepted by Planck, Wien, Sommerfeld, and others. Einstein's advocacy of duality was even seen as a paradoxical retreat from the idea of lightquanta of 1905, back toward the classical view. Even in 1912, Planck's main argument was that the quantization should apply to the action in an atomic process rather than to the energy of emitted or absorbed radiation. Similarly, Bohr (1913) did not accept this discreteness as strictly applying to the radiation. Wave–particleduality was once again a subject of discussion at the Third Solvay Conference in 1921.

De Broglie (1922) made the first attempt to study black–bodyradiation in the context of lightquanta (Medicus, 1974, p. 38). Later, this interest in the properties of quanta motivated de Broglie's search for a theory that would unify the wave and particle aspects. In 1923, he published three short articles in *Comptes Rendus* in which he generalized this wave–particleduality to include material corpuscles (de Broglie, 1923a). This was followed by a short note to *Nature* (de Broglie, 1923b) and then the complete article in *Philosophical Magazine* (de Broglie, 1924).

De Broglie's First Attempt to Postulate Wave-Particle Duality

De Broglie (1923a) applied the wave–particleduality hypothesis to existing problems in physics and among others he referred to the following important issues (cf. Medicus, 1974, p. 40):

a. Application of his hypothesis to electron orbits in an atom, requiring that the wave be in phase with itself, and that the circumference be an integral multiple of the wavelength. In the summary of his article he concluded: "We believe that this is the first physically plausible explanation for the Bohr–Sommerfeld stability rules" (Reproduced in Medicus, 1974, p. 40).
b. The famous formula, $\lambda = h/mv$, is found in this explicit form for the first time in the chapter on statistical mechanics. For de Broglie it is not the wavelength of the particle but the frequency that is more important.
c. There is no essential difference between photons and particles. De Broglie assumed that lightquanta have an extremely small mass of the order of 10^{-50} g.
d. De Broglie even suggested a possible experimental confirmation of his hypothesis in the following terms: "A beam of electrons passing through a very small opening could present diffraction phenomena. This is perhaps the direction in which one may search for an experimental confirmation of our ideas" (reproduced in Medicus, 1974, p. 40). The note to *Nature* (de Broglie, 1923b) also referred to this possibility.

Finally, de Broglie (1924) in a seminal paper explored the reconciliation of lightquanta (used by Einstein to explain the photoelectric effect) with, "the strong experimental evidence on which was based the wave theory" (p. 446).

Experimental Evidence to Support de Broglie's Theory

It is interesting to note that de Broglie's theory of wave–particleduality not only preceded its experimental confirmation, but he also suggested how experiments could be performed. In 1968, in an interview with F. Kubli (1970), de Broglie said that his brother Maurice de Broglie had suggested that the theory should also include an experimental part. De Broglie declined, saying he was not an experimentalist. During his doctoral examination in November 1924 at the Sorbonne, J. Perrin, chairman of the examination committee, asked how one could experimentally observe matter waves (cf. Medicus, 1974, p. 40). De Broglie once again suggested diffraction experiments on crystals with electrons. In a letter written to H.A. Medicus on 17 September 1973, de Broglie said that at the time of his thesis he had suggested to A. Dauvillier (an astrophysicist) to undertake experiments to obtain diffraction or interference phenomena with electrons. Due to experimental difficulties, these efforts were unsuccessful. Furthermore, in a letter written to H.A. Medicus on 16 November 1973, Dauvillier reported: "it is indicative that neither M[aurice] de Broglie, nor P. Langevin, nor J. Perrin saw to it that [such experiments] were carried out in their laboratories! Nobody believed in it" (reproduced in Medicus, 1974, p. 40).

Davisson and Germer (1927) provided the first experimental evidence that helped to understand de Broglie's theory. Interpretation of the data, however, was not straightforward. A scientific meeting in Oxford in August 1926 helped Davisson to consult M. Born and other physicists, who brought him onto the right track (Medicus, 1974, pp. 43–44). Davisson and Germer made a preliminary announcement of their results in *Nature*, April 16, 1927 and then published the complete report in *Physical Review* in December, 1927 and concluded:

> Because of these similarities between the scattering of electrons by the crystal and the scattering of waves by three– and two–dimensional gratings, a description of the occurrence and behavior of the electron diffraction beams in terms of the scattering of an equivalent wave radiation by the atoms of the crystal, and its subsequent interference, is not only possible, but most simple and natural. This involves the association of a wave–length with the incident electron beam, and this wave–lengthturns out to be in acceptable agreement with the value h/mv of the undulatory mechanics. *(Davisson & Germer, 1927, p. 707)*

Einstein's Support of de Broglie's Ideas

Contrary to popular belief, de Broglie was not a young, unknown student when he wrote his major paper on wave–particleduality in 1923. His earlier work led him to various controversies (especially Bohr's Copenhagen school and Sommerfeld's Munich school) that provided resistance to the acceptance of his novel wave–particleduality hypothesis (for details see Raman & Forman, 1969). Despite the controversy surrounding de Broglie's work, Einstein was a strong supporter of wave–particleduality from the very beginning. Langevin (de Broglie's thesis advisor) sent a copy of the thesis to Einstein before the examination (November, 1924). Einstein wrote back to Langevin saying de Broglie had, "lifted a corner of the great veil" (reproduced in Medicus, 1974, p. 41). In December 1924, Einstein was working on a paper in which he calculated the fluctuations of an ideal gas (Bose–Einsteinstatistics) and he incorporated de Broglie's ideas, (published in the *Proceedings of the Prussian Academy* in February 1925, Einstein, 1925), with respect to the association of a wave field with a material particle. Einstein showed how de Broglie's $m_0c^2 = h\nu$ necessarily leads to a wave, exposes the relation between phase and group velocity, and the fact that the group velocity of the wave is the velocity of the particle. Furthermore, he drew attention to an important geometrical interpretation of the Bohr–Sommerfeld quantization rule. Einstein's (1925) support and use of de Broglie's ideas influenced at least two other physicists: Walter Elsasser in Göttingen and Erwin Schrödinger in Zurich (p. 41, Medicus, 1974). In his paper on gas theory, Schrödinger (1926a) noted: "This means nothing else but to take seriously the de Broglie–Einsteinundulation theory of moving corpuscles" (p. 95). Later in his major paper on the relationship between wave and matrix mechanics, Schrödinger (1926b) explicitly acknowledged that his theory was inspired by de Broglie's ideas. These events set the stage for Schrödinger to develop de Broglie's ideas (cf. Darrigol, 1993).

Why Was it Schrödinger who Developed de Broglie's Ideas?

De Broglie's thesis on wave–particle duality was given to Schrödinger by V. Henri, but returned shortly with the remark: "That's rubbish" (reproduced in Raman & Forman, 1969, p. 311, also Jammer, 1966, p. 258). Langevin, on knowing this, instructed Henri to insist that Schrödinger have another look at the thesis. According to Raman and Forman (1969): "even Schrödinger, the man who was in some way fated to develop de Broglie's ideas into a wave mechanics, did not recognize these ideas as promising when first exposed to them" (p. 311).

It is plausible to suggest that Schrödinger was particularly suited to develop de Broglie's ideas for the following reasons: (a) Schrödinger, a marginal man, a loner, was not a member of the Copenhagen/Munich schools of spectroscopists, and still made important contributions to theoretical spectroscopy; (b) Einstein's early endorsement of de Broglie's theory was crucial for Schrödinger, which he acknowledged explicitly in a letter to Einstein dated April 23, 1926: "Your and Planck's assent are more valuable to me than that of half the world. Moreover, the whole business [wave mechanics] would certainly not yet, and perhaps never, have been developed (I mean not by me) if the importance of de Broglie's ideas hadn't been put right under my nose by your second paper on gas degeneracy" (reproduced in Raman & Forman, 1969, p. 311; second paper on gas degeneracy refers to Einstein, 1925); (c) Schrödinger (1922) had previously published ideas which were conceptually similar (although not identical) to de Broglie's (cf. Raman & Forman, 1969, p. 310); (d) Schrödinger's involvement in the fundamental problems of quantum statistical mechanics (Klein, 1964); (e) Conceptually, Schrödinger aligned himself with Einstein in his controversy with the Copenhagen school of quantum mechanics on important issues such as causality, probability, and determinism.

Based on a historical reconstruction it is plausible to elaborate the following aspects of de Broglie's research program, namely 'science in the making':

De Broglie's theoretical framework (presupposition): Plausibility of a reconciliation of Einstein's lightquanta used for explaining the photoelectric effect with classical wave theory of light (de Broglie, 1924, p. 446).

Research questions: (a) Why do electrons in the Bohr atom move in circular orbits at fixed distances from the nucleus? (b) How can the hypothesis be applied to electron orbits in an atom in order to provide plausible explanation for the Bohr–Sommerfeld rules? and (c) What is the significance of the formula, $\lambda = h/mv$, which provides the wavelength of the particle.

Operationalizing heuristic principle: If all waves have properties of particles then all matter has properties of waves. Based on this de Broglie even suggested how this heuristic principle could be operationalized: "A beam of electrons passing through a very small opening could present diffraction phenomena. This is perhaps the direction in which one may search for an experimental confirmation of our ideas" (de Broglie, 1923, reproduced in Medicus, 1974, p. 40).

Designing experiments: Based on the original ideas of de Broglie, Davisson and Germer (1927) designed the experiments and provided evidence for wave–particle duality in the diffraction of electrons by crystals.

Understanding the nature of science: A physics textbook author presented de Broglie's dilemma in the following terms:

> In 1923 Louis de Broglie, then a graduate student, introduced the conjecture that particle–likeobjects (say electrons) should display wave properties . . . Wavelike objects show particle properties (that is, light is emitted and absorbed very much as a particle). This was demonstrated by Planck and Einstein and was utilized by Bohr in his atom. Why not, then, those objects that we normally conceive of as particles? (Let us say electrons.) Why should they not display wave properties? Indeed, why not? What circular orbits were for Plato, harmonies among whole numbers for Pythago–

> ras, regular solids for Kepler, or the solar system centered about the giver of light for Copernicus, this symmetry between wave and particles was for de Broglie. *(Cooper, 1970, p. 341)*

No wonder, Leon Cooper (Nobel laureate in Physics, for details see Chapter 6), once again has provided a historical framework in order to understand de Broglie's "conjecture." In order to accomplish this Cooper takes us through a historical journey from Pythagoras (580–500 bc), to Plato (428–348 bc), to Copernicus (1473–1543) to Kepler (1571–1630). This historical sweep ranging over 2,000 years provides us with an overview of how scientists try to work from guiding assumptions based on conjectures to the actual resolution of a problem. De Broglie's work was questioned by some of his contemporaries and still within the historical perspective, his contribution seems to be within bounds of how original ideas are conceived.

Of the 114 general chemistry textbooks analyzed (published in the USA, Marcano, 2010), 24 (21%) presented de Broglie's dilemma and its possible experimental resolution within a historical perspective. Following are two examples from general chemistry textbooks that clearly conceptualized that de Broglie had suggested wave–particleduality (1923) before experimental evidence and at the same time anticipated the experiments based on diffraction of crystals:

> In his theory of the photoelectric effect, Einstein showed that light has particle–likeas well as wave–likeproperties. In 1923, the young French physicist Louis de Broglie posed the question: If light, which we think of as a wave, can have particle–likeproperties, then why cannot particles of matter have wave–likeproperties? He demonstrated that Bohr's assumption about the quantization of angular momentum in the hydrogen atom could be rationalized by ascribing such wave–likeproperties to the electron . . . In 1927, C. Davisson and L.H. Germer showed that electrons in fact were diffracted by a crystal. De Broglie's relationship correctly predicted their wavelength. The experiment provided a brilliant confirmation of de Broglie's hypothesis about the wave–likenature of matter. *(Oxtoby, Nachtrieb, & Freeman, 1990, pp. 684–686)*

> Then, in the early 1920s, a young French physics student named Louis de Broglie proposed a startling reason for fixed energy levels [Bohr's orbits]: *if energy is particle-like, perhaps matter is wave-like* . . . De Broglie reasoned that *if electrons have wavelike motion* and are restricted to orbits of fixed radii, that would explain why they have only certain possible frequencies and energies . . . If particles travel in waves, electrons should exhibit diffraction and interference . . . In 1927, C. Davisson and L. Germer guided a beam of electrons at a nickel crystal and obtained a diffraction pattern. *(Silberberg, 2000, pp. 273–274)*

These presentations clearly show how in scientific development, some scientists can be intrigued by certain phenomena and, based on their previous experience and presuppositions, postulate research questions that can help to resolve the dilemma. Some scientists go beyond by suggesting heuristic principles, namely operationalizing the possible ways of understanding a problem and its possible solution. The next step is the designing of the experiment itself with all its difficulties and feasibility. Examples from general chemistry textbooks (and Cooper) presented here provide a fairly good picture of what were the problems related to the conceptualization of wave–particle duality and how they were resolved. However, these textbooks and others in the study completely ignored the fact that it was very difficult to design experiments based on diffraction by crystals and, subsequently, to interpret the results. De Broglie alludes to these difficulties and his experience with some experimental scientists who were reluctant to undertake the experiments (for details, see Niaz, 2009a, Chapter 12). Walter Elsasser at the University of Göttingen was one these scientists. Influenced by the ideas of both Einstein and de Broglie he suggested some experiments and discussed them with Einstein, who told him: "Young man, you are sitting on a gold mine!" (reproduced in Medicus, 1974, p. 43). Furthermore, Davisson and Germer (1927) themselves refer to how difficult it was to design the experiments and, later, how interpretation of the data was even more difficult.

At this stage it is important to note that of all the textbooks analyzed (Marcano, 2010) none mentioned how the scientific community itself questioned de Broglie's novel hypothesis of wave–particle duality. Both, Bohr and Sommerfeld, two of the important quantum physicists of the early twentieth century had rejected de Broglie's ideas. This aspect is important in order to understand how scientists themselves subject their work to peer evaluation and thus obtain feedback. Of course, this evaluation in itself leads to conflicts and controversies among scientists, as the different parties try to solicit support from the scientific community. Some textbooks completely ignored the controversial nature of de Broglie's hypothesis and the following is an example:

> Before 1900, it was supposed that light had wave nature. However, the work of Planck, Einstein and others suggested that light behaved as if it consisted of particles, called photons. In 20 years, the dual character of light was accepted almost generally. In 1924 a young French physicist Louis de Broglie came with a revolutionary idea with respect to the nature of matter. Reasoning like the physicists do, at times with great success, de Broglie suggested that particles can exhibit wave properties. *(Masterton, Slowinski, & Stanitski, 1986, p. 176)*

Such presentation in textbooks is a cause of concern for two reasons: (a) It suggests that starting from 1900, in the next 20 years, almost all physicists had accepted the dual character of light. This statement is not supported by the historical record, as the controversy went well beyond 1924, after de Broglie's "revolutionary idea"; (b) It implies that de Broglie was successful as he was "reasoning like the physicists do," that is, coming up with revolutionary ideas. This completely ignores the role played by the scientific community and how in the case of wave–particleduality it led to considerable controversy.

Some textbooks simplified the issues by suggesting that at the time de Broglie suggested his novel idea there was no experimental evidence to substantiate his hypothesis. For example: "When de Broglie put forward his hypothesis, there were no experimental observations to support it, and it was not taken seriously" (Gillespie, Humphreys, Baird, & Robinson, 1989, p. 346). Such statements are somewhat misleading as they seem to suggest that any hypothesis that does not have experimental evidence in support has no role to play in scientific development.

Another group of textbooks referred to the wave–particleduality in the context of a contradiction between the new and the existing dominant paradigm and the following is an example:

> At the beginning of the decade of 1920 there were two characteristics that were particularly objectionable in the field of theoretical physics. One was the conflict between the wave and photon models of light. The other was that the idea of quantized energy had to be *imposed* on Newtonian mechanics, almost like a late reflection. It appeared that it was necessary to establish a new mechanics, in order to soften the wave–particleconflict and introduce the concept of quantized energy as a consequence of certain more basic principles. *(Mahan & Myers, 1987, p. 464, italics in original)*

Such presentations come quite close to the idea of conflicts and controversies in scientific development due to the different presuppositions and guiding assumptions of the scientists working in a field of study. Nevertheless, it still lacks a more explicit reference to the difficulties faced by de Broglie's model of wave–particle duality.

It is plausible to suggest that the historical reconstruction of the wave–particle duality presented in this chapter can have implications for textbooks and teaching of this topic in introductory physical science courses, by emphasizing the following NOS aspects: (a) Based on their presuppositions scientists can postulate novel ideas in order to resolve existing problems; (b) New ideas based on a particular theoretical perspective (presuppositions) inevitably lead to conflicts and controversies with scientists holding the existing dominant research program; (c) As the controversy proceeds, scientists try to find experimental means to provide support for the novel ideas; (d) Experimental data are generally difficult to produce and, in turn, involve a considerable amount of discussion based on alternative interpretations; (e) Postulation of new hypothetical ideas for which there is no existing experimental evidence is a perfectly acceptable research strategy in the history of science.

Conclusions and Educational Implications

A reconstruction of historical episodes based on presuppositions, research questions, heuristic principles and designing experiments, namely 'science in the making' facilitated an understanding of the following aspects of nature of science:

a. Experimental data do not necessarily and in a straightforward manner lead to the formulation of scientific theories and laws. Examples: Law of multiple proportions in chemistry; Millikan and Ehrenhaft's interpretations of data to determine the elementary electrical charge; Davisson and Germer's (1927) interpretation of diffraction of electrons by crystals.
b. A theoretical framework based on a scientist's presuppositions has some speculative aspects. Examples: Maxwell's kinetic theory of gases, Bohr's model of the atom, De Broglie's postulation of wave–particleduality.
c. Presuppositions of a scientist can be based on inconsistent foundations. Examples: Maxwell's kinetic theory of gases; Bohr's model of the atom.
d. Alternative interpretations of experimental data are controversial and at the same time provide a greater understanding of scientific progress. Examples: Accommodation and prediction in the development of the periodic table; explanation of periodicity in the periodic table as a function of an inductive generalization and/or atomic theory; Lenard's trigger hypothesis, Planck's quantum analog, and Einstein's quantum hypothesis to explain photoelectric effect; based on similar data, Millikan postulated the electron and Ehrenhaft the sub–electron.
e. The tentative nature of scientific theories. Examples: Early attempts to understand the periodic table were also based on theoretical frameworks just as the modern atomic law was based on atomic numbers; atomic models of Thomson, Rutherford, Bohr, and Bohr–Sommerfeld; varying interpretations of the photoelectric effect. Interestingly, de Berg (2010) refers to this aspect of the NOS as "discipline in transit" (p. 864).
f. Scientists often proceed by speculating, hypothesizing, proposing models, offering explanations, and formulating a theory, which contrasts with the traditional view of the scientific method. Examples: Thomson's model of the atom; de Broglie's wave–particleduality.
g. A theory is more of a policy rather than a creed for suggesting, stimulating, and directing experiments for charting unexplored areas of study (suggested by Thomson, 1907, in the context of his atomic model).
h. Difficulties involved in the design of an experiment lie as much in developing necessary techniques as in its conceptualization, namely, heuristic principles (suggested by Cooper, 1992, in the context of Thomson's experiments).
i. Scientists generally are reluctant to give up and change their presuppositions (hard–coreof beliefs): Examples: Thomson's reluctance to give up his own model and accept Rutherford's nuclear atom; in the Millikan–Ehrenhaft

controversy, the latter did not give up his presupposition of fractional charges until 1941 (cf. Ehrenhaft, 1941); Millikan's reluctance to give up the classical wave theory of light and later acceptance of Einstein's lightquanta to explain the photoelectric effect.

j. Postulation and acceptance of a new model is difficult due to the rivalry with the existing models. Example: Difficulties and uncertainty faced by Rutherford before postulating his nuclear atom; de Broglie's postulation of wave–particle duality.
k. Conjugation of *experiments* and *conjectures*, namely, the heuristic principles help scientists to understand and conceptualize experimental data (based on Cooper, 1992, in the context of Rutherford's postulation of the nuclear atom).
l. Scientific theories at times do not solve all the problems but, rather, create new dilemmas. Example: Bohr's model of the atom; Bohr–Sommerfeld model of the atom (cf. Niaz & Cardellini, 2011); Perl's muon–electronpuzzle to muon–electron–taupuzzle.
m. In the early periods of theory construction, a scientist needs to hold in abeyance final judgments, lest a promising theory be discarded (based on Holton, 1978, in the context of the oil drop experiment). Example: Millikan's reluctance to accept data that did not provide evidence for the existence of the electron.
n. Doing experiments means gathering data and its interpretation, that is, scientific judgment. Without such judgments journals would be full of mistakes and some scientists can be misled in their judgments (suggested by Olenick et al., 1985 in the context of Millikan's experiments).
o. Science is self–correctingas another scientist with a different heuristic principle can present an alternative interpretation (suggested by Olenick et al., 1985 in the context of Millikan's experiments).
p. There are no dispassionate, unbiased observations as suggested by the scientific method, that is, observations are theory–laden(suggested by Olenick et al., 1985). Examples: Alternative interpretation of the photoelectric effect based on the theoretical frameworks of Lenard, Planck, Millikan and Einstein; Millikan's interpretation of data from his oil drop experiments.
q. Underdetermination of scientific theories by experimental evidence. Examples: Millikan's experimental data in the photoelectric experiments did not necessarily prove Einstein' theory but, rather, provided support for his equation.
r. As the controversy proceeds, scientists try to find experimental means to provide support for the novel ideas. Examples: Bohr's interest in experimental data from atomic spectra; de Broglie's own suggestions for doing experiments and later experimental evidence provided by Davisson and Germer (1927).
s. Postulation of new hypothetical ideas for which there is no existing experimental evidence is a perfectly acceptable research strategy in the history of science. Examples: Bohr's model of the atom; de Broglie's wave–particleduality.

It is important to note that these conclusions are based on the actual historical context in which hypothetical ideas were postulated and experiments conducted. In most cases the format suggested by the original scientist/author has been conserved in order to provide a more authentic understanding of the contribution of different protagonists. This precisely constitutes 'science in the making' and shows the need for teaching science as practiced by scientists (Niaz, 2010a).

From 'Science in the Making' to Contextual Teaching (Science Stories)

Previous sections of this chapter have shown how 'science in the making' in a historical context facilitates an understanding of the NOS. The next step would be to incorporate these historical episodes in the context of the science curriculum and elaborate appropriate, unfolding science stories (Stinner, 1995). Klassen (2006) has referred to this contextual approach in cogent terms:

> School science lacks the vitality of investigation, discovery, and creative invention that often accompanies *science-in-the-making* . . . The humanizing and clarifying influence of history of science brings the science to life and enables the student to construct relationships that would have been impossible in the traditional decontextualized manner in which science has been taught. *(p. 48, italics added)*

An effective way in which to bridge the gap between school science, and what scientists actually do, that is, '*science in the making*' is through the inclusion of humanizing aspects of the history of science in the form of a story (Stinner, McMillan, Metz, Jilek, & Klassen, 2003). Egan's "narrative understanding" would also approximate to this contextual approach:

> students in senior and college courses are drowning in specialized content—in a sea of facts, descriptions, laws and equations. Egan insists that some sort of *metanarrative* that provides an overall explanatory framework of scientific knowledge, change and advance should be presented to students at some point in their science education at the upper level. *(Schulz, 2009b, p. 266, original italics)*

Indeed, for most science curricula and textbooks, science is considered to be a presentation of "facts, descriptions, laws and equations" with no reference to the underlying heuristic principles. Interestingly, both general chemistry and physics textbooks almost completely ignore these heuristic principles (Niaz, 2009a) which constitute the cornerstone of the "structure of the discipline" (Schwab, 1962). Without going to extremes, what we need to do is *weave the narrative around the heuristic principles*, lest we throw away the baby with the bath water. In other words, there are heuristic principles to be drawn upon and a story to be told. It is a cause for concern that most of our textbooks and classroom practice, lack both of these emphases and present science as a "rhetoric of conclusions" (Schwab, 1962). Actually, the situation is worse, as some textbooks do not even comply with the traditional curriculum (Schulz, 2009a, p. 228), by providing appropriate technical details. At this stage it is important to note that the contextual approach need not endorse Egan's (1983) "psychological fallacy" according to which the Piagetian stage theory of development could be in conflict with an alternative fully developed educational theory (cf. Galagovsky, 2011; Schulz, 2009a).

The role of history, besides conceptual understanding is to introduce the humanistic element which provides students the opportunity to become effectively involved with the vicissitudes faced by the scientists themselves. Cushing (1978) would endorse such an approach in order to show, "science to be a human endeavor" (p. 114). Metz (2007) has explored the relationship between science stories based on the HPS and the implemented curriculum. Klassen (2006) refers to this as the *Story-Driven Contextual Approach*, SDCA, that depends on five interacting contexts: (a) Practical; (b) Theoretical; (c) Social; (d) Historical; and (e) Affective.

An example of this contextual approach is illustrated by Millikan's oil drop experiment for undergraduate physics students (Klassen, 2009). This laboratory experience facilitated the following aspects of students' understanding: (a) Humanizing the role played by Millikan and his graduate student (Harvey Fletcher); (b) Difficulty in obtaining results in speculative cutting-edge experiments (e.g., oil drop), if the traditional scientific method is followed rather than allowing presuppositions to guide interpretation of data; (c) The difficult and frustrating nature of the experiment, which is well documented in physics education research; and (d) In the history of science, the oil drop experiment was fundamental in our understanding of the elementary electrical charge. One of the students provided the following insight after this contextual and novel laboratory experience:

> If we had used the data for every drop we observed, our results would have not agreed with the accepted value at all. I suppose that Millikan must have depended quite heavily on his preconception of the value of *e*, assuming his apparatus was similar to ours. If Millikan did not have a testable basis for rejecting drops . . ., I cannot see how the experiment would give one confidence that charge quantization had been observed. *(reproduced in Klassen, 2009, p. 604)*

The merit of such statements from students needs to be appreciated in the context of research that shows the dilemma faced by most students while interpreting data from the oil drop experiment (cf. Jones, 1995). Furthermore, it bears witness to the effectiveness of the contextual approach based on historical reconstructions. Indeed, Kolstø (2008) has suggested that the Millikan–Ehrenhaft story provides a good argument for inclusion in a science curriculum for citizenship. A recent study has provided further support for innovative teaching strategies (Paraskevopoulou & Koliopoulos, 2011), in the context of the Millikan–Ehrenhaft controversy with respect to the elementary electrical charge.

Another example of the contextual approach is provided by the photoelectric effect and its use in the undergraduate physics courses (Klassen, 2011). Presentation of the photoelectric effect in general physics textbooks and undergraduate laboratory manuals, is associated with various myths with respect to the roles played by Planck, Lenard, Einstein, and Millikan. The contextual approach is based on a story entitled, "Birth of the Photon Concept," integrated with instruction that includes live demonstrations, whole–class concept quizzes, chalkboard illustrations, and worked examples. This classroom experience based on the contextual approach led Klassen (2011) to conclude:

> By simplifying and misrepresenting the nature of scientific discoveries in isolation, we may, in fact, produce the very thing that we, as science teachers, want to avoid. Pretending that the answers to big questions were resolved in an uncomplicated, problem–free progression destroys or prevents the very engagement and questioning of students that we hope to stimulate. *(p. 12)*

The periodic table of the chemical elements has also been the subject of some preliminary attempts to go beyond the traditional teacher–oriented classroom practice (Erduran, 2007). Demircioğlu, Demircioğlu, & Çalik (2009) introduced the periodic table to 9th grade students by embedding storylines based on the context approach. The storylines not only improved the students' ability to link theoretical knowledge with real life but also facilitated their meaningful learning of the periodic table.

Wiebe and Stinner (2010) have suggested a storyline starting from the controversy between the atomists and the Aristotelians up to the postulation of the kinetic molecular theory, as a means to facilitate students' understanding of gas properties. Differentiation between heat energy and temperature by our students (also in the history of science) is facilitated by the conceptual shift from caloric theory to kinetic molecular theory (Wiser, 1988). Niaz (2000c) has reported that freshman students fall back on the caloric theory even after having responded correctly in a context that approximates the kinetic view of heat energy and temperature. Laburú and Niaz (2002) found that secondary school students held alternative models such as, "heat is a form of substance" and "temperature is a form of heat." A conceptual change teaching strategy based on cognitive conflicts facilitated a transitory model. However, students holding the transitory model were not entirely convinced that increase in heat energy does not increase the temperature of boiling water. Some students experienced a further progressive transition by constructing a scientific model based on the understanding that "Temperature only measures the energy of molecular agitation." Interestingly, according to de Berg (2008a), "much of the language of heat even in current textbooks is reminiscent of the old caloric theory" (p. 86). Based on his analysis of general chemistry textbooks and an HPS perspective, de Berg (2008a) presents various scenarios that refer to the following questions: (a) What is the relationship between the amount of heat and the mass of water? (b) What is the relationship between temperature rise and the mass of water? and (c) What is the relationship between temperature rise and the amount of heat? These questions can be incorporated into a storyline in order to facilitate students' understanding of heat energy, temperature, and kinetic theory. In a study conducted by Niaz (2006) to facilitate students' conceptual understanding of heat energy and temperature, a similar strategy was suggested: "What happens when unequal masses of water at different temperatures are mixed?" (p. 274).

3

STUDENTS' AND TEACHERS' UNDERSTANDING OF THE NATURE OF SCIENCE

Introduction

If one could subscribe to Kuhn's thesis of revolutions interspersed with periods of "normal science," science education has indeed gone through a revolutionary period in the last few decades. Delamont (1990) narrates an incident that in retrospect is quite revealing. She relates how at the end of the 1970s a paper which used the theoretical debate between Kuhn and Lakatos as its central "motif" was submitted to an educational research journal. The editor was so perplexed that he exclaimed: "Who on earth are Kuhn and Lakatos?" The interesting point is that the editor was a distinguished professor of science education. Landscape of science education has changed over the last 20 years and researchers have become deeply imbued with various perspectives related to the history and philosophy of science (HPS).

In 1952, Gerald Holton's *Introduction to concepts and theories in physical science*, provided a glimpse for students and teachers as to how science evolves through the interactions of theories, experiments, and actual scientists, within an HPS perspective. Looking back after almost 50 years, Holton (2003) considered that the textbook facilitated understanding of science as a coherent story based on the thoughts and work of living scientists. More recently, a new edition of this textbook has presented science as a human adventure, from Copernicus to Einstein and beyond (Holton & Brush, 2001).

Writing in a special issue dedicated to *Students' Models and Epistemologies of Science*, Linn, Songer and Lewis (1991) pointed out:

> Gerald Holton infused Harvard Project Physics with marvelous historical examples of scientific investigation. These were heralded as ground breaking, but were rarely imitated. Instead, if they discuss process at all, *science*

> *textbooks describe an outmoded and incorrect view of scientific knowledge acquisition. (p. 729, italics added)*

The importance of the HPS was generally recognized by science educators in the 1960s (Klopfer, 1969; Robinson, 1969a, 1969b), and in the last 20 years there has been a worldwide sustained effort to provide a rationale for its inclusion in the science curriculum (Matthews, 1990, 1994a; Scheffler, 1992). Similarly, the objective of helping students to develop informed views of the nature of science (NOS) has been a central goal for science education for the last several decades (Abd-El-Khalick& Lederman, 2000; Abd-El-KhalickBell & Lederman, 1998; Niaz, 2008b; Osborne, Collins, Ratcliffe, Millar, & Duschl, 2003).

Many researchers and teachers are working towards the inclusion of the HPS in the science curriculum, textbooks, and classroom practice (Hodson, 1985, 1988, 2008; Matthews, 1994a; Niaz, 2009a, b; 2011). Interestingly, however, Bevilacqua and Bordoni (1998) have stated: "We are not interested in adding the history of physics to teaching physics, as an optional subject: the history of physics is 'inside' physics" (p. 451). This is an important argument, which has generally been ignored in the science education literature. In other words, we do not have to justify or request for the inclusion of the HPS in the science curriculum and textbooks. Teaching science without the historical context in which ideas and theories develop, comes quite close to what Schwab (1962, 1974) has referred to as "rhetoric of conclusions." Matthews (1998) has argued that philosophy is not far below the surface in any science classroom, as most textbooks and classroom discussions deal among others, with concepts, such as law, theory, model, explanation, cause, hypothesis, confirmation, observation, evidence, and idealization (p. 168). Similarly, Niaz and Rodríguez (2001) based on a historical framework have shown that the HPS is already "inside" chemistry, and we do not need separate courses for its introduction. Abd-El-Khalick(2005) has expressed this idea in cogent terms:

> if we want teachers to address NOS instructionally, our efforts to help them develop the necessary understandings need to go beyond a few hours of NOS-relatedinstruction in a science-methodscourse. Naturally, history and philosophy of science, which are the "stuff" of NOS, are primary candidates for enriching the development of science teachers in the area of NOS. *(p. 38)*

Students frequently believe that science is a collection of facts and that the best way to learn science is to memorize those facts (Linn, Songer, & Lewis, 1991). The degree to which students' conceptions of the NOS are influenced by their teachers and textbooks is the subject of considerable research. According to Lederman (1992), such influence is mediated by a complex set of factors, such as curriculum constraints, administrative policies, and teachers' conceptualization of learning. Developing scientific literacy (as part of the NOS) can be successful only to the extent that science finds a niche in the cognitive and cultural milieu of students (Cobern, Gibson, & Underwood, 1999).

The purpose of this chapter is to review research on students' and teachers' understanding of the NOS within an HPS perspective. Similar reviews, especially with respect to the NOS have been conducted in the past (Lederman, 1992; Meichtry, 1993; Lederman, Wade & Bell, 1998). Later, Bell, Abd-El-Khalick, Lederman, McComas, and Matthews (2001) published a bibliography of research related to the NOS and science education. Consequently, this review has focused on 94 studies published in the period 2004–2008, which provides a wide spectrum of ideas with respect to what is important in areas related to the HPS, and has drawn on the following journals: *International Journal of Science Education*, *Journal of Research in Science Teaching*, and *Science Education*. This is not an exhaustive list of journals of interest to science educators, and some readers may consider the inclusion of *Science & Education* necessary. Nevertheless, it is plausible to suggest that *Science & Education* (a journal devoted to HPS-relatedresearch) would normally cover such studies to a much greater extent. For readers, it would be of more interest to know what is the coverage of HPS-relatedissues in the three major journals, which cover many other areas of interest to science education research? It is important to note that this is primarily a thematic review and it provides graduate students (and others starting a research career in science education) an overview of how the HPS and the NOS are being studied in the science education literature. Furthermore, a critical appraisal of various research categories has also been included. In order to select a study for review, I used the following list of qualitative criteria (research tools): Title of the article, abstract, key words, theoretical rationale, method, conclusion, and references. It is important to note that there are some issues, such as science literacy, science-technology-society (STS), conceptual change, pedagogical content knowledge, analogies, and constructivism, which are important within an HPS perspective. However, articles related to such issues were included only if they had an explicit bearing on some aspect of the HPS. Geographical location of a study is mentioned only if it was conducted outside the USA.

After having selected the studies for review it became apparent that these could be classified in the following two broad categories:

1. Epistemological beliefs of students and teachers with respect to the NOS. Based on the type of study, this category was further subdivided into the following seven sub-categories:

    a. The relationship between students' and teachers' epistemological beliefs.
    b. The myth of the scientific method.
    c. Children's scientific reasoning.
    d. Scientists' views of the NOS.

   e. The NOS and science curriculum.
   f. The NOS and students' laboratory practice.
   g. Science exhibitions as a means to understanding the NOS.

2. Facilitating students' and teachers' understanding of the HPS, based on topics that are already in the science curriculum. Based on the type of study, this category was further subdivided into the following six sub–categories:

   a. The role of argumentation.
   b. Explicit and reflective vs. implicit inquiry–orientedinstruction.
   c. The use of NOS–enrichedinstructional materials.
   d. The use of history–basedinstructional materials.
   e. The use of technology–basedhistorical materials.
   f. The use of science apprenticeship programs.

It is important to note that neither the categories nor the sub–categorieswere generated a priori, but were based on the type of research reported in the three selectedjournals. In other words, issues explored in this chapter emerged from the review of the literature and were not selected by the author.

Epistemological Beliefs of Students and Teachers with Respect to the Nature of Science

Of the 94 studies reviewed, 60 (64%) were classified in this section. Based on the subject, treatment, and orientation of the study, seven categories were generated (see Table 3.1). Studies classified in each of these seven categories are presented here. In some cases, in order to understand the studies included in the review, previous research in the area has also been included.

Relationship between Students' and Teachers' Epistemological Views (n = 27)

Akerson and Abd–El–Khalick(2003) found that most elementary teachers have not heard of the term "nature of science" (NOS) and when they see it in their state frameworks they misinterpret the term as meaning something to do with nature and not with the essence of science itself. Authors suggested that these teachers cannot teach about the NOS without first improving their own views. Some pre–serviceteachers do not develop more informed views of the NOS, despite participation in a science–methodscourse that uses an explicit reflective approach. Furthermore, some teachers do not retain their improved views of the NOS, which shows the difficulties involved in its conceptualization (Akerson, Morrison, & Roth McDuffie, 2006). Southerland, Johnston and Sowell (2006) found that teachers' conceptual ecologies influenced the degree to which they found NOS–basedideas useful. Conceptual ecologies were based among other

TABLE 3.1 Summary of articles on epistemological beliefs of students and teachers with respect to nature of science, n = 60 (published 2004–2008)

Category	*n*	*Articles*
Relationship between students' and teachers' epistemological beliefs	27	Akerson et al. (2006), Chen (2006), Dagher & BouJaoude (2005), Deniz et al. (2008), Dogan & Abd–El–Khalick (2008), Hokayem & BouJaoude (2008), Holbrook & Rannikmae (2007), Irez (2006), Kang et al. (2005), Lawson (2005), Lin & Tsai (2008), Liu & Lederman (2007), Liu & Tsai (2008), Marshall & Young (2006), Park (2006), Roehrig & Luft (2004), Sadler (2004), Sadler & Zeidler (2005), Sandoval (2005), Sensevy et al. (2008), Smith & Wenk (2006), Southerland et al. (2006), Tsai (2007), Tsai & Liu (2005), Yalvac et al. (2007), Yang (2004), Yilmaz–Tuzun& Topcu (2008)
Myth of the scientific method	3	Lin & Chiu (2004), Windschitl (2004), Windschitl et al. (2008)
Children's scientific reasoning	4	Kawasaki et al. (2004), Russ et al. (2008), Tytler & Peterson (2004), Warwick & Siraj–Blatchford(2006)
Scientists' views of the nature of science	9	Coll & Taylor (2004), Ford (2008), Florence & Yore (2004), Samarapungavan et al. (2006), Schwartz & Lederman (2008), Sevian & Gonsalves (2008), Taylor et al. (2008), Yore et al. (2004), Yore et al. (2006)
The nature of science and science curriculum	10	Abd–El–Khalicket al. (2004), Allchin (2004, a, b), Enfield et al. (2008), Hipkins et al. (2005), Matthews (2004), Oulton et al. (2004), Slater (2008), Van Aalsvoort (2004), Westerlund & Fairbanks (2004)
The nature of science and students' laboratory practice	6	Havdala & Ashkenazi (2007), Kang & Wallace (2005), Kittleson & Southerland (2004), O'Neill & Polman (2004), Reiner & Gilbert (2004), Vhurumuku et al. (2006)
Science exhibitions as a means to understanding the nature of science	1	Davidsson & Jakobsson (2007)

factors on epistemological beliefs, conceptions of science, self–efficacy,and religious beliefs. These researchers emphasize the role played by extrarational factors on the development of teachers' views of the NOS. Similarly, Sadler and Zeidler (2005) have reported that science classrooms are environments in which intuition and emotion in addition to reason are valued. In a critical review of research on informal reasoning, Sadler (2004) concluded that socioscientific issues can provide a forum for working on informal reasoning, argumentation skills, NOS conceptualizations, the evaluation of information, and the development of conceptual understanding of science content. Furthermore, Sadler considered that Kuhn (1962) challenged the significance of formal reasoning in science by proposing a novel model of scientific change in which competing theories are judged by a variety of criteria including informal reasoning and social influences.

With respect to the role played by induction and deduction in scientific inquiry, Lawson (2005) has found support for Popper's (1965) hypothesis that enumerative induction does not exist as a psychological process. In contrast, the author suggests that progress in science is characterized by increasingly abstract cycles of hypothetico-deductive reasoning. Consequently, science instruction should provide students with opportunities to generate and test increasingly complex and abstract hypotheses and theories.

Factors that impacted the inquiry-basedinstruction of 14 beginning secondary science teachers were explored by Roehrig and Luft (2004). Teachers were followed for a period of one year, in an attempt to understand their teaching beliefs, instructional practices, knowledge of the NOS and scientific inquiry, and their experiences with inquiry instruction in the classroom. Case and cross-case comparisons revealed five main constraints that impacted their enactment of inquiry-basedinstruction: understanding of the NOS and scientific inquiry, content knowledge, pedagogical content knowledge, teaching beliefs, and concerns about management and students.

According to Sandoval (2005), inquiry-based instruction is generally advocated for developing students' understanding of the NOS, although there is scant evidence that it actually does that. There is a gap between what is known about students' inquiry practices and their epistemological beliefs about science. Studies of students' ideas about epistemological aspects of formal science are unlikely to shed any light on how they perceive their own inquiry efforts. Inquiry-based instruction that does not take into account the epistemological beliefs that guide students' inquiry, stand very little chance of helping students to understand professional science. Sandoval's main argument is that students' beliefs about the NOS and their practices of inquiry are guided by *practical epistemologies*. It is concluded that the study of these practical epistemologies has the potential to produce a better psychological theory of epistemological development and thus realize the goals of science education.

Relations among various aspects of freshman college students' epistemologies of science were explored by Smith and Wenk (2006). At the start of their first semester, 35 college freshmen were interviewed in order to probe: (a) their differentiation of scientists' ideas from evidence, and hypotheses from theories; (b) their understanding of the inherent uncertainty of scientific knowledge; and (c) their reasoning about scientific controversies (one of the controversies was whether fluoridation of water prevents tooth decay). The most common responses were in terms of an epistemology in which scientists' ideas and evidence are differentiated, and theories are understood as tested hypotheses, although students varied in how consistently they differentiated theories and evidence across all questions. Responses in which theories are understood as broader explanatory frameworks guiding hypothesis testing were virtually nonexistent, but some students gave responses that showed awareness of processes of interpreting and reinterpreting patterns of results. Responses across the three parts of the interview were significantly related. Consistently differentiating scientists' ideas from evidence was strongly related to appreciating the inherent uncertainty of scientific knowledge and with having a deeper understanding of the reasons for scientific controversies and how to resolve them (cf. Chapter 2). Authors consider their work to have complemented that of Driver et al. (1996) by making some finer distinctions among students' understanding of the reasons for scientific controversies and in relating this understanding to their emerging domain-specificknowledge of experimental design. Following groups of students were observed: (a) students who simply expected diverse data; (b) students who were beginning to articulate different variables that might affect the results or how well a treatment worked; and (c) students who understood that the scientists could have been entirely mistaken in their interpretations because of uncontrolled confounding variables.

Marshall and Young (2006) have reported a study of three science teachers' development of theories-in-actionas they worked together in a group to explore collisions using both physical manipulatives and a computer simulation (Interactive Physics™). Theories-in-actionas originally proposed by Karmiloff-Smith and Inhelder (1975) are learners' "implicit ideas or changing modes of representation." Analyses based on an existing theoretical framework indicated that, as the group moved from physical experiments to the computer simulation, their attention shifted from planning their experiments to processing system feedback, which impeded the iterative refinement of their theories-in-action. The nature of the theories they developed also changed. Learners' attitudes toward science and prior experiences affected the exploration process in both environments. In particular, prior instruction in physics and an authoritarian view of science seemed to impede engagement in the development and testing of theories-in-action.

The role of theory and evidence in a socioscientific issue (use of underground water) in 10th grade students from Taiwan was studied by Yang (2004). The reasoning mode was assessed using a questionnaire with open-endedquestions. Results showed that students' ability to incorporate theory and evidence was in general inadequate, and their reasoning mode was consistent with their epistemological perspectives. Furthermore, male and female students appear to have different reasoning approaches.

Tsai (2007) has explored the relationship between middle school physical science teachers' (Taiwan) epistemological views, teaching beliefs, and instructional practices, and students' epistemological views. Findings suggested adequate coherence between teachers' epistemological views and teaching beliefs as well as instructional practices. Teachers with relatively positivist-alignedviews tended to draw attention to students' science scores in tests and allocate more instructional time on teacher-directedlectures and in-classexaminations, thus implying more passive or rote learning. In contrast, teachers with constructivist-orientedviews tended to focus on student understanding and application of scientific concepts, by devoting more time to inquiry activities and interactive discussions.

Chen (2006) has reported the development of a valid, meaningful, and practical instrument for creating in-depth profiles of the views of college students or adults, including pre-/in-service teachers, about the NOS, and NOS instruction. The questionnaire, Views on Science and Science Education (VOSE), includes 15 questions, each followed by several items representing different philosophical positions. The instrument was validated with students in Taiwan and the final version was administered to 302 college students. Combined conceptions and conflicting thoughts about the NOS were detected, quite similar to those in the USA. For example, concerning the question about how scientists choose a theory between competing theories, some subjects strongly agreed that scientists accept both theories before they can objectively distinguish the best (competition among rival theories is observed frequently in science) and, on the other hand, also believed that scientists tend to choose the theory which deviates less from the contemporary core theory. The author concluded that the content and format of VOSE were an improvement on the weaknesses of the traditional questionnaires (VNOS and VOSTS).

Structural relationships between scientific epistemological views (SEVs) and information commitments (ICs) were explored by Lin and Tsai (2008). Data were collected from 486 Taiwanese high school students via two self-reporting instruments: (a) an SEV questionnaire, including five scales for representing students' views toward scientific knowledge; and (b) an IC survey, involving six scales exploring evaluative standards and searching strategies of online science information. Structural equation modeling analysis was used to examine the relationships between the various aspects of SEVs and ICs. Results showed that students' SEVs guided their evaluative standards and searching strategy when dealing with web-based science information. For example, students who viewed scientific knowledge as more changeable and tentative significantly tended to adopt a more sophisticated strategy by carefully inspecting the content of web sites for judging the usefulness. This clearly shows the implication of an important NOS aspect (tentativeness of knowledge) in classroom practice. In general, students with more constructivist-oriented SEVs developed more advanced standards and searching strategies toward online scientific information. Authors concluded that the role of SEVs should be highlighted as an increasingly metacognitive engagement with online science information.

Liu and Tsai (2008) have compared the SEVs of science major (pure science and science education) and non-major undergraduates. A multidimensional instrument (developed previously by Tsai & Liu, 2005) was used to assess the SEVs of 220 freshmen (42% science and 58% non-science majors) attending two public universities in Taiwan. Results indicated that the science majors have less sophisticated beliefs in the theory-laden and cultural-dependent aspects of science, than non-science majors. Analysis of variance results further revealed significant differences in SEV dimensions among the three major fields: non-science, pure science and science education. Science education students gained the lowest scores on the entire scale among the three groups. Authors suggested the following plausible explanations for these findings: (a) Science major (including science education) students might have been involved longer in such an epistemic environment that described scientific knowledge as objective and universal; and (b) Possibly, beliefs about certainty and objectivity lead these students to select science as the major field.

The relationship between an individual's culturally based worldviews (humans' understanding of the natural world) and conceptions of the NOS has been explored by Liu and Lederman (2007). Conceptions of the NOS and worldviews of 54 Taiwanese prospective science teachers were assessed using two open-ended questionnaires in conjunction with follow-upinterviews. Understandings of the NOS were classified into informed and naive categories based upon contemporary views of these constructs and those stressed in international reform documents. Findings showed that: (a) Participants who recognized the limitations of scientific knowledge, and accepted the idea that science involves subjective and cultural components were more likely to emphasize harmony with Nature; and (b) Participants who possessed narrow views about the scientific enterprise and described science as close to technology, tended to provide an anthropocentric perspective regarding the human–Nature relationships. These findings illustrate the interplay between participants' sociocultural beliefs and conceptions of the NOS and suggest the need for incorporating sociocultural perspectives and the NOS in the science curriculum.

Turkish prospective science teacher educators' (biology, chemistry, physics) understanding of the NOS has been studied by Irez (2006). A reflection-oriented qualitative approach was adopted and data were collected through two sets of interviews and analyzed by means of cognitive maps. Analysis of the data revealed that most of the participants had inadequate conceptions regarding the NOS. The most problematic aspects were the scientific method and the tentative NOS. Only 20% of the participants agreed that "there is not a single scientific method." Interestingly, although all the participants accepted the tentative NOS theories, still 60% perceived scientific laws as the final form of scientific knowledge, representing truth and hence not subject to change.

Turkish pre-service science teachers' epistemological views have also been explored by Yalvac et al. (2007). Similar to the previous study from Turkey (Irez, 2006), most participants agreed that scientific knowledge is tentative, but still did not fully understand the differences between hypotheses, laws, and theories.

Dogan and Abd-El-Khalick(2008) elucidated high school Turkish students' and science teachers' conceptions of the NOS to help inform curriculum materials development at the high school level in relation to the NOS. The study was based on a representative national sample comprising 2,087 students and 378 science teachers. All participants were administered a questionnaire comprising

14 modified items from the "Views on Science–Technology–Society"(VOSTS), originally developed by Aikenhead and Ryan (1992). The NOS aspects targeted by these 14 items were the following:

a. The theory–driven nature of scientific observations (role of guiding assumptions)
b. The tentative nature of scientific knowledge
c. The relationship between scientific constructs (models) and reality
d. The epistemological status of hypotheses, theories, and laws
e. The nature of and relationship between scientific theories and laws
f. The myth of a universal and/or stepwise "scientific method"
g. The nonlinearity of scientific investigations
h. The role of probabilistic reasoning in the development of scientific knowledge.

This may seem to be an ambitious agenda. However, it is important to note that these and similar aspects of the NOS have been emphasized in science education reform documents (e.g., AAAS, 1993; NRC, 1996) and by science education researchers (Lederman et al., 2002; Niaz, 2008b; Osborne et al., 2003). Dogan and Abd–El–Khalick(2008) based on their study found that a majority of the participants held naive views of most aspects of the NOS. Interestingly, teachers' views of the NOS were mostly similar to those of their students, which is an important finding for the implementation of new teaching strategies.

Yilmaz–Tuzunand Topcu (2008) have explored Turkish pre–serviceelementary science teachers' (PSTs) epistemological beliefs and the relationships among their epistemological beliefs, epistemological worldviews, and self–efficacy beliefs. Four hundred and twenty–nine PSTs who were enrolled in five large universities completed the Schommer Epistemological Questionnaire (SEQ), the Epistemological World Views Scale, and the Science Teaching Efficacy Belief Instrument. Factor analysis results revealed four factors for the SEQ. These factors were Innate Ability, Simple Knowledge, Certain Knowledge, and Omniscient Authority. Multiple regression analysis suggested, for "Innate Ability" factor scores, that three of the predictor variables—self–efficacy,outcome expectancy, and worldview—contributed significantly to the model. For "Simple Knowledge," only one predictor variable—epistemological worldview—contributed significantly to the model. For "Certain Knowledge" factor scores, only one predictor variable—outcome expectancy—contributed significantly to the model. None of the predictor variables significantly contributed to the "Omniscient Authority" factor scores. Results revealed that in Turkish culture, PSTs' epistemological beliefs support the multidimensional theory. In addition, while PSTs developed more sophisticated beliefs in some of the SEQ dimensions, they had less sophisticated beliefs in other dimensions.

Deniz, Donnelly and Yilmaz (2008) have explored the factors related to acceptance of evolutionary theory among pre-service Turkish biology teachers (n = 132) using conceptual ecology for biological evolution as a theoretical lens. In Turkey, the Qur'an and its interpretations heavily influence public understanding of how the universe, the earth, and living things came into existence. The teleological clash between Darwinian evolution and the Qur'anic verses is a major obstacle that causes many religious people to reject evolution as a scientifically valid theory. The purpose of the study was to account for the variance in acceptance of evolutionary theory based on the following factors: Understanding of evolutionary theory, epistemological beliefs, thinking dispositions, and parents' educational level as independent variables. Authors expected that participants with relativistic-subjective views of knowledge would be more likely to accept evolutionary theory than participants with dualistic-objectivist views of knowledge. Results showed that: (a) Pre-servicebiology teachers' thinking dispositions, their understanding of evolutionary theory, and their parents' educational level are positively correlated with acceptance of evolutionary theory; and (b) There is no significant positive correlation between epistemological beliefs and acceptance of evolution. Together these factors explained 10.5% of the variance. These results suggest that studying the relationship between acceptance of evolutionary theory (a controversial issue) and other related factors in a multivariate context is more informative. Authors advocate a constructivist teaching strategy emphasizing the constructed nature of knowledge as opposed to the notion that knowledge of evolution is a direct representation of reality.

Kang, Scharmann and Noh (2005) have examined Korean students' views on the NOS. An empirically derived multiple-choiceformat questionnaire was administered to 1,702 Korean 6th, 8th, and 10th graders. The questionnaire consisted of five items that respectively examined students' views on five constructs concerning the NOS: the purpose of science, the definition of scientific theory, the nature of models, the tentativeness of scientific theory, and the origin of scientific theory. Students were also asked to respond to an accompanying open-ended section for each item in order to collect information about the rationale(s) for their choices. According to the authors, such a format has advantages over a traditional Likert-typeinstrument and/or an interview in the case of a large-scalesurvey. The results indicated that the majority of Korean students possessed an absolutist/empiricist perspective about the NOS. It was also found that, on the whole, there were no clear differences in the distributions of 6th, 8th, and 10th graders' views on the NOS. In some questions, distinct differences between Korean students and those of Western countries were found. Authors recommended that it may be more productive to teach students about the NOS at the elementary school than to remedy secondary students' inadequate understanding.

Generating an explanatory hypothesis to explain a discrepant event is important for students' conceptual change. Park (2006) has investigated how undergraduate Korean students generate new explanatory hypotheses. As a first step, the author identified students' prior ideas related to electromagnetic induction. After showing six conflicting phenomena, six students were asked to suggest explanatory hypotheses to explain the phenomena. Using interviews, the processes of generating explanatory hypotheses were analyzed and three types were identified: theoretical, experiential, and auxiliary hypotheses. It is concluded that subjects use similarity–basedreasoning to relate their background knowledge or experiences with the conflicting phenomena to be explained.

Holbrook and Rannikmae (2007) have explored the meaning of NOS education to enhance scientific literacy. A model based on activity theory rather than logical positivism, is presented. This encompasses understanding based on cooperative learning and socioscientific decision–making.

An epistemological approach to modeling based on the work of Cartwright, Fleck and Hacking has been developed by Sensevy et al. (2008). This leads to emphasizing the transitions between the abstract and the concrete in the modeling process. Four case studies illustrate how students in France are engaged in constructing relations between the abstract and the concrete through modeling activities, by elaborating at the same time specific language games and appropriate thought styles. It is suggested that embedding knowledge on the one hand, and classes as thought collectives, on the other hand, may contribute to understanding science education.

College students' understanding of the nature of the theory of evolution and its scientific status has been explored by Dagher and BouJaoude (2005). Authors conducted semi–structuredinterviews with 15 college biology seniors, who were asked to explain why they thought evolution assumes the status of a scientific theory, how it compared to other scientific theories, and what criteria they used to determine if an explanation is scientific or not. Students' responses encompassed five themes that included: evidence, certainty, experimentation, method of theory generation, and prediction. These themes focused on the theory's empirical dimension which seemed to be derived from a generic and simplistic model of physical science theories that valued direct empirical evidence. Demanding that evolutionary theory conform to this model reveals a misunderstanding of its nature. This misunderstanding was expressed in relation to aspects of methodology, explanation, and prediction. Findings underscore the need for using explicit discipline–and context–specificapproaches to teaching and learning about scientific theories.

Although a well–corroboratedscientific theory, the theory of evolution continues to cause dilemmas for some students who have not easily been able to accommodate the concepts of this theory within their "cognitive culture." Perhaps the reason lies in the overlap of some ideas that the theory advocates with other social, epistemological, and religious beliefs. Hokayem and BouJaoude (2008) have investigated how 11 Lebanese college biology students, who completed a course on the theory of evolution, perceive the relationship among their epistemological beliefs about science, their beliefs about religion, and their position regarding the theory of evolution. Questionnaires and semi–structuredinterviews were used to collect data. Qualitative methods were used to analyze the data and identify the various positions of the students and the course instructor. Following is an example of a response from a student who rejected evolutionary theory:

> They haven't scanned the whole earth to see if what they're talking about is true, it's fragmented, they find one thing they make up a theory on it, they find something else, they change their theory . . . they're basing it on their imagination nothing else . . . they need to show me transitional species that are really found, not just they found a human being with a bigger jaw, it's normal for the jaw to be bigger because he used to eat other kind of food, if this is what they mean by evolution then fine but not the evolution from monkeys to human, this is another idea. *(reproduced in Hokayem & BouJaoude, 2008, p. 407)*

Such ideas can be found among both students and teachers in many parts of the world. For example, the widespread belief that the "whole earth needs to be scanned," "find one thing they make up a theory on it," and "they find something else, they change their theory," can be found among various cultures and in different areas of knowledge. Interestingly, however counter-intuitive these beliefs may appear, this is how science progresses, namely scientists do not necessarily have to go around the world collecting experimental data before postulating a theory, new findings can change an existing theory and this precisely shows the tentative nature of scientific knowledge (see Chapter 2 for other examples). While discussing religion and science, this student considered his religious beliefs very strong and thought that religion played an important role in his life because it organized his life from a practical and moral aspect. The positions of the students in this study were similar to those of Lebanese students in a previous study (Dagher & BouJaoude, 1997) who had not taken a course in evolution. The argument regarding the nature of evidence was crucial for accepting the theory or not accepting it in both studies. One difference, however, was that the student who rejected evolution in this study did not claim to do so due to a commitment to the religious text as the ones in Dagher and BouJaoude's study claimed. Another difference between this and the previous study was that none of the students in this study were "noncommitted," refusing to take a stance regarding the theory. It seemed that the advantage of spending a few months in studying evolution did not change the arguments that defended the students' positions but it made them restructure those arguments in such a way that fit within their worldview in light of new acquired knowledge.

Studies reviewed in this section reveal that both students and teachers have considerable difficulty in understanding the NOS, which shows the complexity of the issues involved as teachers must understand the NOS in order to teach. In general there is coherence between teachers' epistemological views and instructional practice. Most students understand science in the traditional empiricist sense that is absolute and objective with the scientific laws as the final form of scientific knowledge. This makes any change difficult, especially with respect to understanding theories as broader explanatory frameworks guiding hypothesis testing and based on arguments and evidence, rather than direct representation of reality.

The Myth of the Scientific Method (n = 3)

The National Society for the Study of Education (1947) emphasized the scientific method for teachers in the following steps: making observations, defining the problem, constructing hypotheses, experimenting, compiling results, and drawing conclusions. This oversimplified view of what constitutes the scientific endeavor has proven to be resistant to change and is used in almost all parts of the world. Lederman et al. (2002) have clearly traced its origin to Francis Bacon's *Novum Organum*, and its unhealthy influence in science education.

According to Grandy and Duschl (2007) the logical empiricist conception of science is closely related to the traditional scientific method, namely: make observations, formulate a hypothesis, deduce consequences from the hypothesis, make observations to test the consequences, and accept or reject the hypothesis based on the observations (p. 150). Despite all the changes in modern philosophy of science, school science and textbooks still espouse the traditional scientific method. Grandy and Duschl (2007) have suggested that the theory-drivenand model-drivenviews of science can help to change and expand the traditional scientific method.

Windschitl (2004) reported a study in which 14 pre-servicesecondary science teachers developed their own empirical investigations, from formulating questions to defending results in front of peers. Findings indicated that the teachers consistently used the following misrepresentations of some fundamental aspects of science: (a) A hypothesis functions as a guess about an outcome, but is not necessarily part of a larger explanatory system; (b) Background knowledge may be used to provide ideas about what to study, but this knowledge is not in the form of a theory or model; (c) A theory is an optional tool one might use at the end of a study to help explain results. Finally, Windschitl (2004) concluded that these ideas appear consistent with a "folk theory" of an atheoretical scientific method, "*that is promoted subtly, but pervasively, in textbooks, through the media, and by members of the science education community themselves*" (p. 481, emphasis added).

Lin and Chiu (2004) examined the relationship between student understanding of the NOS and their problem-solvingstrategies. Six hundred and twenty, year 8 students in Taiwan completed two conceptual problem-solvingtests and a questionnaire on the NOS. Results of the stepwise multiple regressions indicated that the subscale of the scientific method consistently appeared as the best predictor for student problem-solvingability, explaining about 22% of the variance. It was also found that problem-solvingstrategies were more conceptually based for students who were high scorers on the NOS survey.

More recently, Jenkins (2007) has traced the origin of the scientific method as a political construct in the nineteenth century, and that it is at odds with developments in the HPS in the twentieth century. According to Windschitl, Thompson and Braaten (2008), 100 years after its conception, the scientific method continues to reinforce a kind of cultural lore about what it means to participate in inquiry. Windschitl (2004) has referred to the crux of the issue by pointing out that textbooks and science educators themselves seem to espouse the scientific method. The textbook by Olenick et al. (1985) provides guidelines for science educators by pointing out that there are no dispassionate, unbiased observations as suggested by the scientific method (see Chapter 2 for details).

Children's Scientific Reasoning (n = 4)

Research on children's scientific reasoning has been inspired by Piaget's seminal work (Inhelder & Piaget, 1958), which led him to conclude that the logic of scientific experimentation and inference is not acquired until adolescence. Metz (1997), however, has argued that Piaget's work is based on active manipulation of variables and experimental controls required to support the construction and elaboration of causal theories. Kuhn (1997), in contrast, has presented evidence to support that children tend not to think of theory and evidence as separate entities. Sodian, Zaitchik and Carey (1991) presented evidence that 1st and 2nd grade students can indeed distinguish belief from evidence in situations that do not require covariation arguments, and do not contradict prior beliefs. Niaz (1997) has critiqued the methodology used by Sodian et al. (1991) on the grounds that it constitutes a "training" study. Koslowski (1996) has added a new ingredient to the debate by arguing that much work on scientific reasoning, where children are asked to make decisions about causal relationship on the basis of correlational evidence, represents an empiricist view of the scientific method. Given the considerable difficulty faced by even adolescents in hypothetico-deductivereasoning (Adey & Shayer, 1994), the subject of how early children understand some form of scientific reasoning remains a controversial issue (cf. Biao et al., 2009; Niaz, 1997).

An alternative to understanding students' views of the NOS and epistemological reasoning has been presented by Driver, Leach, Millar and Scott (1996):

a. Phenomenon-basedreasoning, in which explanation and description are not distinguished, and the purpose of experimentation is to observe
b. Relation-basedreasoning, in which explanation is presented in terms of relations between observable entities
c. Model-basedreasoning, in which theories or models are evaluated in the light of evidence, and the relationship, is recognized as provisional and problematic.

Later, Leach, Driver, Millar and Scott (1997) presented age-related characterizations of students' scientific knowledge. Tytler and Peterson (2004) have extended the work of Driver et al. (1996) by characterizing different dimensions and levels of scientific reasoning in early elementary school children, in the context of open explorations. The study is based on Australian children's (age, about 5 years) responses to three tasks during two interviews in which they explored three science phenomena: behavior of mealworms, what determines whether objects float or sink, and the flight of paper whirlybirds. Authors were particularly interested in the nature and level of children's scientific reasoning, the extent to which they could generate explanations, and what evidence would be required to distinguish between competing knowledge claims. Results obtained showed that: (a) Children who were operating in hypothesis-checkingmode in the floating and sinking, and whirlybird tasks, were clearly coordinating theory and evidence at a higher level, considered to be concept-basedreasoning; (b) In all these cases children were using or generating a hypothesis and devising a way to produce evidence to test this; (c) Capacity to articulate and explicitly shape the coordination of ideas and evidence was tentative and uncertain in every case; (d) This illustrates the distinction between modes of epistemological reasoning and ability to manage forms of scientific exploration, such as control of variables.

Evolution of children's (3rd/4th grade) theory building and modeling as critical components of scientific epistemologies, over a unit of study on sinking and floating, was examined by Kawasaki et al. (2004). The study employs discourse analytic methods in order to examine epistemological development in the dynamic swirl of talk, and social interaction in the classroom, based on the work of Driver et al. (1996). Based on changes in student participation and analyzing the quality and content of these discussions, authors concluded that they are developing a powerful epistemology of science that is more commensurate with the work of actual scientists.

Within the HPS, the "scientific revolution" has been characterized as a shift from "occult" to mechanistic accounts of the natural world (Westfall, 1986). Science education, research, and curriculum development have focused on promoting mechanistic reasoning over teleological and anthropomorphic reasoning for understanding. In this respect, a mechanistic idea is more "scientific" than an occult or teleological idea, whether or not it is correct. Russ et al. (2008) have developed a framework for discourse analysis that aids in identifying and analyzing students' mechanistic reasoning (based on the work of Machamer, Darden, & Craver, 2000). Based on this framework the authors analyzed a discussion among 1st grade children about falling objects to: (a) Illustrate the generativity of the framework, and (b) Demonstrate that mechanistic reasoning is abundantly present even in young children.

The development of a science education that includes a focus upon the NOS suggests the need for "pedagogical tools" that can be used to engage children with the procedural understandings that are central to the scientific approach to enquiry. Warwick and Siraj–Blatchford(2006) have reported a collaborative action research project that focused on the use of secondary data as just such a "tool" for stimulating engagement with procedural understanding among primary school children in the UK. These authors suggest that the comparative analysis of secondary and investigative data can provide a basis for such engagement, provided children have guided opportunities to discuss their understanding of the issues revealed by the comparisons.

Research reported in this section shows that an incipient form of model–based reasoning is used even by young children. Interestingly, given an appropriate classroom environment, children can even demonstrate mechanistic reasoning (in contrast to "occult" ideas) to understand the natural world of falling objects.

Scientists' Views of the Nature of Science (n = 9)

Hogan and Maglienti (2001) have compared the ecological understanding of science professionals (scientists and technicians) and non–scientists(adults and middle school students). Quantitative and qualitative analyses revealed that responses of the two groups differed especially with respect to their emphasis on criteria of empirical consistency and plausibility of conclusions. Interestingly, the authors attribute this difference not so much to developmental capabilities but, rather, cultural preferences. When asked to elaborate with respect to this preference all of them mentioned that they learned to judge conclusions based on evidence through the *social processes of doing science* and following is an example of one of the participating scientists' responses:

> We get it from the community, by seeing how people react to things, how they judge things, what the standards are. When you are around it enough, you internalize the rules. It is not like law school where there are probably strict rules laid down and you learn them. Nobody teaches us these things or stands up at a blackboard and lists them, but we all know how to do it. *(reproduced in Hogan & Maglienti, 2001, p. 679)*

Indeed, this is quite revealing. In other words, "to judge conclusions based on evidence," that is, interpretation of the data, is difficult and not based on strict rules (scientific method) but, rather, requires the participation of the scientific community. Such experiences can inform efforts in science education to engage students in actual scientific practice. According to Niaz (2009a), interpretation of experimental data is difficult and inevitably leads to alternative models/theories thus facilitating understanding of 'science in the making' as a human enterprise.

Yore, Hand and Florence (2004) have emphasized that an understanding of scientists' ontological and epistemic beliefs about science is essential to understand their scientific practice, especially with respect to doing, writing, and reviewing research. Among others, these authors formulated the following research questions: (a) What views of the NOS are held and used by academic researchers?; and (b) Are academic researchers' beliefs about the NOS reflected in their beliefs about writing strategies and processes? The study is based on 19 university faculty members, working in science-relateddisciplines (Biology, Biochemistry, Microbiology, Chemistry, Computer science, Earth and Ocean sciences and Astronomy) at a midsize Canadian university. All participants were interviewed and responded to a science writer questionnaire and asked to select one of the following images that represented their understanding of the NOS:

> Image A (Absolutist view)
> This view of science suggests that science knowledge is developed through observations, measurements, and plausible reasoning. Facts are accumulated by making observations and measurements; and these data are generalized to form a big idea . . . Scientific knowledge, according to the absolutist view, is a collection of truths that is unchanging. *(Yore et al., 2004, p. 348)*
>
> Image B (Evaluativist view)
> This view of science proposes that science knowledge is a temporary explanation that best fits the existing evidence and current thinking. Science knowledge claims develop with the aid of a hypothesis (causal statement) and data that are collected either to support or refute the hypothesis. Hypotheticodeductive reasoning relies on the absence of refuting evidence and the presence of confirming evidence as support for a hypothesis. Since scientists have preconceptions about the outcome of an investigation, most evidence the scientist would likely collect would support the existing hypothesis rather than refute the hypothesis. *(Yore et al., 2004, pp. 349–350)*
>
> Image C (Relativist view)
> This view of science knowledge proposes that people construct explanations in the context of their own personal beliefs. The different interpretations and explanations of the natural world are expected and considered of equal value. The relativist view of science does not question, evaluate, or judge divergent interpretations and explanations, because one explanation cannot be judged as more valid than another explanation because different interpretative frameworks were used. *(Yore et al., 2004, p. 350)*

Based on the data obtained from the interviews and the questionnaire, it was found that all of the scientists positioned themselves around the modern, evaluativist view of science, while rejecting the extremes of the traditional, absolutist and postmodern, relativist view. These scientists described the evaluativist nature of science by using terms and phrases such as debate, argument, hypothesis testing, tentative, best temporary idea, and moving toward more accurate perceptions. One of the scientists expressed this in the following terms:

> One has to listen to the new information. We have a concept of nature, a model or a theory, and we are trying to understand whether the world around us is consistent with that theory . . . sometimes we find our understanding has to be modified in accord with the world around us. The judge of validity is independent of human constructions. *(reproduced in Yore et al., 2004, p. 352)*

Yore et al. (2004) consider such understanding of the NOS as illustrative of a naive realist ontological view of the world. This makes interesting reading and contrasts with the realist view of the NOS (close to the absolutist view in this study), that is, reality can be discovered and accurately described by human activities and these mental representations will correspond with the material objects in the real world (Bickard, 1997; Staver, 1998). On the other hand, naive realists assume that there is a real world beyond sensory perception, although it may be impossible or at least difficult to disentangle the descriptions of this reality from human perceptions and limitations. These are difficult questions (important nevertheless), and most scientists, textbook authors, and teachers do not engage in this debate. In this study, most of the scientists expressed views that came close to the evaluativist (naive realist) view, but still expressed some doubts and reservations, for example:

> I think there is a truth perhaps, but maybe a changing truth . . . I guess my main point of view would be that anything you see written in science is the best we can do at this point. But it's all a hypothesis. There probably is an answer, but we don't have the means to know for sure we've got it. *(reproduced in Yore et al., 2004, p. 352)*

This study also found that the internal consistency of the scientists' responses about writing and specific language choice did not completely align with their views of science (Research question b). For example, they reported results in a language that did not fully match the epistemology inherent in the naive realist view of science in which interpretation of data goes beyond simple generalizations from experimental data. Interestingly, some of the participants perceived a conflict with respect to reporting research in an active or passive voice. One scientist expressed the view that writing in the passive voice can be used dishonestly. Another expressed the view that the use of entirely passive voice was "outmoded" and favored the mixing of passive voice (methods section) and active voice (discussion section). Finally, Yore et al. (2004) concluded: "This movement toward the use of active voice reflects greater alignment with the contemporary view of the nature of science and potentially recognizes the human dimension in data interpretation and knowledge construction" (p. 362).

Florence and Yore (2004) explored how graduate students learn to write like scientists and the ensuing enculturation into the processes that characterize 'science in the making.' These novice scientists came to appreciate that the writing, editing, and revising process influenced the quality and understanding of science. It also became apparent how peer-reviewedjournals are the dominant gatekeepers of the scientific research communities. One of the findings of the study involved the incongruent relationship between the scientists' use of language and their evaluativist views of science, and the sociocultural dynamics of research groups seeking consensus in their deliberations. Finally, these authors concluded, "These scientists appeared to believe that the science community's gatekeepers, traditions, and conventions that are influenced by an absolutist view of science, dictate the language, voice, and style of reporting science" (p. 662). These are important findings and clearly show the role played by the scientific community (peer review) in the construction of scientific knowledge. The degree to which this role converts the scientific community into "gatekeepers" is, of course, a cause of concern.

Yore, Florence, Pearson, and Weaver (2006) have reported an autobiographical case study of two scientists who had participated in previous studies. The central issue of interest was: Whether these modern, evaluativist scientists believed that the review-react-reviseprocess of publishing a peer-reviewedresearch report simply improves the quality of the language or actually changes the science. The scientists served both as informants and co-authors.Both scientists believed that writing and revising research reports improved the science as well as the clarity of text and that their use of absolutist language related to their beliefs about inquiry and not about science knowledge.

The research question, "Just how open-mindedare modern scientists?" was investigated by Coll and Taylor (2004). The Exeter questionnaire to examine superstitious beliefs of high school students and pre-servicescience teachers was used as a basis for a series of in-depth interviews of scientists across a variety of disciplines in New Zealand and U.K. universities. The interviews sought to understand the basis on which scientists form beliefs and how they judge evidence for various propositions and other contentious beliefs introduced during discourse. The scientists are dismissive of traditional superstitions such as bad luck associated with black cats and inauspicious numbers such as 13. Testimony of others including media reports are deemed unreliable unless provided by credible witnesses such as "educated people" or "experts," or if they coincide with the scientists' personal beliefs. These scientists needed potential theoretical explanations for beliefs and were generally dismissive of empirical evidence without underlying explanations.

Schwartz and Lederman (2008) examined practicing scientists' views of the NOS and explored possible relationships between these views and science context. Science educators emphasize teaching the NOS through inquiry-based learning experiences throughout all science disciplines. Yet aspects of the NOS that are agreed upon as relevant to science education have been described in discipline-independentways. Given the situated nature of learning, these authors sought to examine how scientists understand recommended aspects of the NOS and whether these views vary depending on the science discipline or investigative context of authentic practice. Participants were 24 scientists, representing four broad science disciplines (chemistry, life science, physics, earth/space science) based upon a variety of approaches (experimental, descriptive, theoretical). Data included questionnaires and interviews. Through intra-group and cross-group comparisons, associations between views of the NOS, scientific disciplines, and methods of scientific inquiry were examined. Results obtained indicated that views of the NOS are not necessarily related to science context. There appears to be as much variation within groups as across groups. Differences in views seem to be embedded within individual contexts and experiences rather than broader science disciplines. A majority of the scientists acknowledged the influence of current scientific theory and paradigm in directing scientific research. They recognized theory-ladenobservations and investigations from within their research contexts and provided examples. Authors considered this view more sophisticated in comparison with only recognizing personal subjectivity, such as variances in taking measurements. Interestingly, few beginning science teachers achieve this deeper level of understanding (cf. Schwartz et al., 2004). Overall, the participants in this study affirmed that scientific knowledge is subject to change and recognized there are areas of science that are more certain than others, yet some viewed science as progressing toward knowledge of external reality. Results support the applicability of advocated NOS aspects across science contexts, and do not support a need to teach different natures of science in the K-12science arena.

Taylor, Gail Jones, Broadwell, and Oppewal (2008) explored the perceptions of scientists from diverse science domains and science teachers, through semi-structured interviews. Participating scientists expressed concern about the variability in the quality of teaching, programs, and resources available for science instruction. Scientists expressed a desire for teachers to have more experience conducting science research, to develop their own critical thinking skills, and to make science exciting for their students.

Ford (2008) has explored the relation between how scientific knowledge is created and the reasoning involved in learning content with understanding. His study offers a relation that draws coherent connections between the science studies literature, which suggests ways of conceiving how scientific knowledge is created, and sociocultural learning theory, which suggests ways of conceiving scientific reasoning. This model highlights the dialectic between *construction* and *critique* of claims in both scientific reasoning and practice. A "grasp" of scientific practice as such is instrumental to learning because informational content of scientific knowledge lies not only on the level of facts, but also on the levels of methods and values, and coordinating information across these levels is crucial for understanding. In contrast to prevailing constructivist ideas that highlight student *authority* to construct knowledge as scientists do, this model emphasizes the importance of knowing how to hold claims *accountable*. Thus, the ideal vision of students making *their own sense* of content is superseded by a more defensible ideal vision of students learning how to make *scientific sense* of content. Given the present state of understanding constructivism in science education, this study provides insight and a possible alternative for going further (cf. Niaz, 2008c).

Samarapungavan, Westby, and Bodner (2006) have investigated the ways in which beliefs about the NOS vary as a function of an individual's chemistry expertise and chemistry research experience across the range from high school students, whose exposure to chemistry occurs in the classroom, to practicing research chemists. Interviews conducted with a total of 91 participants probed three key research questions: Do the participant's epistemic beliefs vary as a function of chemistry expertise? Are there discipline–specificvalues and heuristics that guide chemistry research? How does research experience influence participants' epistemic beliefs. Differences in both the duration and the nature of participation in research had a significant effect on how participants conceptualized science and scientific research. Only the practicing chemists saw a productive role for empirical anomalies that arise in the course of doing research, and they conceptualized scientific work in terms of a building or "engineering" model of science. This, contrasts with the classic hypothetico–deductivemodel of science invoked by some science educators (cf. Lawson, 2003). Authors concluded that current characterizations of the NOS in science education may underrepresent important discipline–specificaspects of science.

A rubric for assessing the quality of scientific explanations provided by science graduate students has been developed by Sevian and Gonsalves (2008). The rubric was developed from a qualitative analysis of science graduate students' abilities to explain their own research to an audience of non–scientists.It helps to evaluate scientific explanations by distinguishing between the content knowledge and pedagogical content knowledge of scientists, as well as a scientist's ability to integrate the two in the service of a clear and coherent explanation of research.

Research reported in this section reveals that practicing scientists consider the writing and revising of research reports as an important part of the scientific endeavor, and that peer–reviewedjournals are the dominant gatekeepers of the scientific research communities. Generally, scientists recognized theory–laden observations and that views of the NOS are not necessarily related to science context.

The Nature of Science and Science Curriculum (n = 10)

Slater (2008) has raised a provocative question for science education: How can teaching false science be justified? This, in turn, is based on the premise that we teach false science. Among others, Slater discusses the case of Newtonian mechanics and Bohr's model of the atom—both are false and still we teach them to countless high school and college students around the world. This may sound preposterous to students and even perhaps teachers. According to Slater (2008):

> It seems to me that the "business–as–usual,'textbook–sciencemodel of science education faces two closely related worries. On the one hand, teaching as true what we or our students may otherwise reasonably regard as strictly false theories may feed a similar misconception amongst present–day science students (and adults alike) that science is *done*, that we know pretty much everything about the world, and that what is left is stamp collecting and number crunching. *(p. 540)*

Interestingly, both Newtonian mechanics and Bohr's model of the atom are good examples for illustrating the tentative nature of scientific knowledge (cf. Niaz, 2009a, 2011). Despite the predominance of Newtonian mechanics for over 200 years, at the beginning of the twentieth century physicists found alternative interpretations based on relativity theory and quantum mechanics (Giere, 2006). Similarly, in the case of atomic models, the change has been even more explicit and rapid, based on the atomic models of Thomson, Rutherford, Bohr, Bohr–Sommerfeld and wave mechanics (for details, see Niaz & Cardellini, 2011). In this historical context, it would be pertinent to ask if Newtonian mechanics and Bohr's model of the atom constitute "false science." Interestingly, science curricula in general and textbooks simply ignore such issues. An HPS–basedperspective would suggest that scientific models are not "false" or "true," but vary in the degree to which these can explain or predict experimental findings.

According to Van Aalsvoort (2004) most secondary school students consider chemistry to be irrelevant. Based on a critical review of science education literature (e.g., Burbules and Linn, 1991), the atheoretical nature of the observational language and the curriculum (based heavily on the textbooks), the author concluded that chemical education is driven by logical positivism. As a philosophy of science, logical positivism creates a divide between science and society. Based on these premises, the author hypothesized that the adoption of logical positivism causes chemistry's lack of relevance in chemical education. This hypothesis was substantiated by an analysis of the secondary school chemistry curriculum in the Netherlands. Based on these considerations, the author concluded:

> Chemical education is relevant from a social point of view to the extent that the knowledge it provides is applicable to solve society's problems. Yet, due to the hierarchical relation between scientific knowledge and its applications, the former is preferred above the latter in chemical education, thereby leaving the relevance of chemical knowledge for society mostly out of sight. *(p. 1166)*

Finally, the author suggested that as an alternative to logical positivism, science educators could explore activity theory within the sociocultural approach.

Various aspects of science education, such as conceptual change, constructivism, and multiculturalism, have been deeply influenced by Kuhn (1962).

According to Matthews (2004), Kuhn is correct to assert that criteria of rationality have to take account of the history of science and that methodological decisions are learned in the practice of inquiry as well as factual lessons. Furthermore, we learn about rationality a posteriori, not just a priori. Nevertheless, the influence of Kuhn has been so pervasive that Loving and Cobern (2000) consider that the science education community has turned into an "admiration society for Thomas Kuhn" (p. 199). Matthews (2004) has cautioned that uncritical acceptance of Kuhn's ideas is not conducive towards the understanding of ongoing debates in the HPS. Looking back, he has argued cogently with respect to how the same lesson was learned from the science education community's wholesale embrace of logical empiricism during the 1950s and 1960s.

The relationship between inquiry learning, science curriculum and the NOS has been envisioned by Abd-El-Khalicket al. (2004) in the following terms:

> Students learn to do inquiry in the context of science content and develop epistemological understandings about NOS and the development of scientific knowledge, as well as relevant inquiry skills (e.g., identifying problems, generating research questions, designing and conducting investigations, and formulating, communicating, and defending hypotheses, models, and explanations). *(p. 398)*

Science has a role to play in the resolution of many of the issues deemed controversial in all societies. The potential role of controversies to understand an important aspect of the NOS has generally been underemphasized or completely ignored. Oulton, Dillon, and Grace (2004) have used the outbreak of "foot and mouth disease" in livestock on English farms in 2001, to illustrate the importance of controversies for the science curriculum. The outbreak led to a controversy about whether to use vaccination, with scientists and scientific evidence being used to support both sides of the argument. Authors suggested that the science curriculum take into consideration the following guidelines: (a) Scientists analyze a problem based on a particular worldview and thereby arrive at different perspectives; (b) Differing worldviews can occur because scientists adhere to different value systems; and (c) Controversial issues cannot always be resolved by recourse to reason, logic or experiment.

The need to include history of science in the curriculum is not free of controversy. Allchin (2003) has shown how some science educators frame historical stories (e.g., Gregor Mendel, William Harvey) leading to a rhetorical architecture of myth that misleads students about how science derives its authority. Westerlund and Fairbanks (2004) agree with Allchin that scientific myths distort history and foster unwarranted stereotypes about the nature of science. However, they point out that although heroic depictions of Mendel often detract from his real contributions, so do claims that discredit them. In his response, Allchin (2004a) suggested that we need to place Mendel in historical context, approaching history "scientifically," independently of preoccupation about "credit" and values (p. 760). In a similar vein, Allchin (2004b) has suggested that science educators need to differentiate between the normative and descriptive elements of nature science, which requires historical reconstructions of 'science in the making.'

The mismatch between curriculum reform rhetoric in science education and actual classroom practice has been explored by Hipkins et al. (2005). These authors suggest the following factors for this mismatch: (a) Lack of philosophical consensus about NOS; (b) Lack of appropriate curriculum guidance, classroom materials, and pedagogical content knowledge; (c) Teachers' personal theories of learning; and (d) Realities of classroom constraints. It is concluded that a wider debate in the community and support are necessary.

Enfield, Smith, and Grueber (2008) have reported on a study of curriculum materials development and use compared with the use of existing curriculum materials in an elementary classroom. The research explored the effect of explicit attention to epistemic practices in curriculum materials and the enactment of those materials, such as: asking questions, collecting data, making descriptions of observations, finding patterns in the data, and developing scientific reasoning. Scaffolding considered the function, role, and purpose of epistemic practices in the community of science and the production of scientific knowledge. The designed curriculum was compared with an existing pedagogically similar curriculum to consider whether the scaffolding and explicitness affected enactment. Findings show that scaffolding in materials created more instances of students engaging in scientific reasoning. Similarly, the explicit description of epistemic practices supported the teacher in explicitly providing opportunities for children to learn about and use epistemic practices of science.

Research reported in this section shows the close relationship between the NOS and science curriculum. Teaching science as true/false may deprive students of the understanding that scientific theories are tentative in nature. For most science textbooks, and even teachers, J.J. Thomson's model of the atom became "false" when E. Rutherford proposed his model of the atom. Teaching history of science inevitably involves such issues and, consequently, an alternative approach would be to consider the heuristic/explanatory power of a theory, instead of the true/false dichotomy (Lakatos, 1970). Similarly, uncritical acceptance of Kuhn's ideas is not conducive towards the understanding of current debates both in science and the HPS. Furthermore, experimental data are difficult to interpret which inevitably leads to controversies that require historical reconstructions of 'science in the making.'

The Nature of Science and Students' Laboratory Practice (n = 6)

In recent years, a number of curriculum reform projects have advocated the notion of having students do science in ways that move beyond hands-onwork with authentic materials and methods. These reformers have argued that students should come to an understanding of science through doing the discipline, that is, 'science in the making.' This stance has occasionally been mocked by its critics as an attempt to create "little scientists," an objective that is either romantic or without purpose. According to the U.S. *National Science Education Standards* (NRC, 1996), one way students will develop scientific ways of thinking is by developing the abilities necessary to: (a) Identify questions and concepts that guide scientific investigations; (b) Design and conduct scientific investigations; (c) Use technology and mathematics to improve investigations and communications; (d) Formulate and revise scientific explanations and models using logic and evidence; (e) Recognize and analyze alternative explanations and models; (f) Communicate and defend a scientific argument (Content Standard A, Grades 9–12, pp. 175–176).

O'Neill and Polman (2004) have presented evidence that educating students in the "little scientists" mode can be a practical way to achieve scientific literacy goals: "Our work suggests that struggling (and even failing) to formulate and carry out empirical investigations in science may teach students more meaningful lessons about how science is accomplished than flawlessly executing cookbook labs or solving carefully formulated problems" (p. 262). These authors report the experience of one student (Beth) in their study who chose to investigate underground nuclear testing, and soon encountered similarly competing claims from environmental organizations and the French government over the issue of whether such testing causes geologic damage. After doing further research on underground nuclear testing, Beth suggested that the environmental organizations were making catastrophic claims without any data to support their conclusions.

Kittleson and Southerland (2004) studied a group of students in a mechanical engineering senior capstone design course to document the interaction patterns and knowledge construction activities. A total of 20 lab sessions were audio taped, in which at least two students were working on a simulation or experiment. Data were analyzed using Gee's (1999) method of discourse analysis. In one of the sessions a group investigated automobile defrosters, which have been used for many years, yet little is known about the heat transfer and fluid dynamics phenomena related to them. Most of the time in lab (both simulation and experimentation) was spent on collecting various types of data. Members of the group believed their data reflected what was actually happening in the real world. In an interview, one of the students reported, "I mean, obviously the experiments are right because, well, they're the real stuff" (p. 284). According to the authors, such views are based on the assumption that one can determine what is really happening, and are congruent with objectivist epistemology and realist ontology. Furthermore, it is not just contextual features such as objects and other group participants that shape knowledge, but also those seemingly invisible factors such as assumptions and ideologies (Kuhn, 1970; Longino, 1990) that each participant or community holds that can affect the knowledge produced.

The relationship between secondary science teachers' epistemological beliefs and teaching goals to their use of lab activities has been explored by Kang and Wallace (2005). Three experienced secondary science teachers were interviewed and observed throughout an academic course. The findings illustrate that a teacher's naive epistemological beliefs are clearly reflected in their teaching practices. Furthermore, various syntheses of different aspects of epistemological beliefs and instructional goals are linked to teachers' diverse ways of using lab activities.

Epistemological roots of knowledge when students carry out hands-onexperiments in physics have been identified by Reiner and Gilbert (2004). Within the context of designing a solution to a stated problem, Israeli students constructed and ran thought experiments intertwined within the processes of conducting physical experiments. Authors suggested that the process of alternating between these two modes—empirically experimenting and experimenting in thought—leads towards a convergence on scientifically acceptable concepts, referred to as mutual projection. In this process external representations were generated. Objects in the physical environment were represented in an imaginary world and these representations were associated with processes in the physical world. It is through this coupling that constituents of both the imaginary world and the physical world gain meaning.

In traditional laboratory work students are, to a large extent, occupied by doing measurements and manipulating apparatus. This does not provide them an opportunity to verbalize theoretical knowledge and relate theory to practice. Havdala and Ashkenazi (2007) designed a laboratory study in which they took the emphasis away from Israeli freshman chemistry students' hands-onwork and, instead, focused on pre-and post-labactivities, which demand consideration of both theory and practice. Students were interviewed with respect to their views about science and the following is an example of an interchange:

Q: How would you define a scientific law?
A: It is something experimental, for sure. It is not God-given,but it is something you can trust to be correct in 99%.
Q: What is it based upon?
A: Only on experiment. (reproduced in Havdala & Ashkenazi, 2007, p. 1141)

Based on the data from the interviews, students were classified into the following groups: empiricist-oriented,rationalist-oriented,and constructivist-oriented. Students' views about science were correlated with their approaches to lab practice. A coherent epistemological theory was constructed for each case, by considering the different degrees of certainty and confidence each student attributed to theoretical versus experimental knowledge in science.

Vhurumuku, Holtman, Mikalsen, and Kolsto (2006) investigated the proximal and distal images of the NOS that A-levelstudents (n = 72) from Zimbabwe developed from their participation in chemistry lab work. Authors also explored the nature of the interactions among the students' proximal and distal images of the NOS and students' participation in lab work. Students' views of the NOS and the nature of their chemistry lab work were elicited through their responses to an open–endedquestionnaire and semi–structuredinterviews. The results suggest that students build some understandings of the NOS from their participation in lab work. Students' proximal understandings of the NOS appear to build into and interact with their understandings of the nature and practice of professional science. This interaction appears to be mediated by the nature of instruction. It is posited that each student's conceptual ecological system is replete with interac–tions which govern attenuation of proximal understandings into distal images. Methodologically, the study illustrates how students' lab work–basedproximal and distal images of the NOS can be identified and extracted through analyzing and interpreting their responses to protocols.

Research reviewed in this section reveals that lab activities are important as they provide an opportunity of understanding science through the discipline that is 'science in the making.' Again, the struggle and the effort involved in doing empirical research are more important than executing flawless cookbook labs, despite the belief to the contrary. Doing experimental work inevitably involves implicit guiding assumptions and epistemological frameworks that are at times ignored by the students. Besides the hands–on activities in the lab, pre– and post–lab activities provide good opportunities for understanding the scientific enterprise. Promoting explicit NOS instruction in the lab, however, remains a challenge and requires the support of policy–and decision–makers.

Science Exhibitions as a Means to Understanding the Nature of Science (n = 1)

Science centers aim to present science in ways that will attract visitors and enhance public interest in, and knowledge of, science. Davidsson and Jakobsson (2007) have investigated the different aspects of science presented by Nordic science centers. Staff members who worked with the planning and creation of new exhi–bitions were asked to answer a web–basedquestionnaire, identifying the extent to which different aspects of science were displayed in their latest exhibition. Findings showed that exhibitions chose to display the wonders of science, by presenting it in a product–orientedand unproblematic way. Participants were also asked their opinion with respect to displaying different aspects in future exhibi–tions. Most participants expressed a desire to emphasize the social and cultural aspects of science.

Facilitating Students' and Teachers' Understanding of the History and Philosophy of Science

Of the 94 studies reviewed, 34 (36%) were classified in this section. Based on the subject, treatment, and orientation of the study, six categories were generated (see Table 3.2). Studies classified in each of these six categories are presented here.

According to Bencze and Hodson (1999) in many cases the science curriculum itself acts as a barrier to change in teaching and understanding science. These authors refer to the following myths with respect to scientific inquiry in school science curricula in Ontario (Canada) and elsewhere: Observation provides direct and reliable access to secure knowledge, science starts with observation, science proceeds via induction, experiments are decisive, science comprises discrete generic processes, science is a value-freeactivity, and scientific inquiry is a simple, algorithmic procedure (p. 522).

Role of Argumentation (n = 9)

Research in science education has recognized the importance of interactive argumentation within the classroom as a means to enhance students' interest, motivation, and conceptual understanding (Boulter & Gilbert, 1995; Chinn & Anderson, 1998; Driver, Newton, & Osborne, 2000; Duschl & Osborne, 2002; Erduran & Jimènez-Aleixandre, 2008; Kelly, Drucker, & Chen, 1998; Kuhn, 1993; Lawson, 2003; Martins, Mortimer, Osborne, Tsatsarelis, & Jímenez-Aleixandre, 2001; Mortimer & Machado, 2000; Newton, Driver, & Osborne, 1999;

TABLE 3.2 Summary of articles on facilitating students' and teachers' understanding of the history and philosophy of science, n = 34 (published 2004–2008)

Category	*n*	*Articles*
Role of argumentation	9	Bricker & Bell (2008), Chang & Chiu (2008), Clark & Sampson (2007), Erduran et al. (2004), Nussbaum et al. (2008), Osborne et al. (2004), Sampson & Clark (2008), Simon et al. (2006), von Aufschnaiter et al. (2008)
Explicit and reflective vs. implicit inquiry oriented instruction	11	Abd-El-Khalick& Akerson (2004), Abd-El-Khalick& Akerson (2007), Akerson et al. (2008), Akerson & Hanuscin (2007), Akerson & Volrich (2006), Hanuscin et al. (2006), Johnston et al. (2006), Khishfe (2008), Khishfe & Lederman (2006), Khishfe & Lederman (2007), Sowell et al. (2007)
Use of NOS-enriched instructional materials	7	Bartholomew et al. (2004), Dagher et al. (2004), Ford & Wargo (2007), Martín-Díaz(2006), Sadler et al. (2004), Volkman and Zgagacz (2004), Wong et al. (2008)
Use of history-based instructional materials	3	Abd-El-Khalick(2005), Hosson & Kaminski (2007), Pocoví (2007)
Use of technology-based historical materials	2	Crawford et al. (2005), Sandoval & Reiser (2004)
Use of science apprenticeship programs	2	Bencze & Elshof (2004), Schwartz et al. (2004)

Niaz, Aguilera, Maza, & Liendo, 2002; Simon, Osborne, & Erduran, 2003; Solomon, 1998; Zohar & Nemet, 2002). Chinn and Anderson (1998) conceptualize *interactive argumentation* as:

> Discussions in which participants present reasons and evidence for different positions. These discussions usually take the form of a conversation rather than a formal debate . . . [and] are rarely hostile or disputatious. Rather, the students generally treat each other respectfully, and their argumentation represents a collective search for reasons and evidence that sometimes leads them to change their minds. *(p. 317)*

A major problem confronting the development of argumentation in the science classroom is that it is primarily a dialogic event carried out among two or more individuals, in order to achieve consensus. Given the authoritarian and dogmatic orientation of most classroom environments, opportunities for dialogic discourse are minimized.

Use of argumentation can also help to introduce the historical context in which a particular topic developed. Niaz et al. (2002) hypothesized that classroom discussions based on arguments/counterarguments used by J.J. Thomson, E. Rutherford, and N. Bohr to elaborate their atomic models, can facilitate students' conceptual understanding. Results obtained showed that given the opportunity to argue and discuss, students' understanding can go beyond the simple regurgitation of experimental details. Performance of the experimental group showed contradictions, resistances, and finally progressive conceptual change with considerable and consistent improvement. The dilemma faced by a science teacher is that a simple reproduction of the experimental details, without the historical context (conflicts and controversies in the interpretation of data), inevitably leads to "rhetoric of conclusions" (Schwab, 1962). Newton, Driver, and Osborne (1999) have expressed this tension between observations, experiments, and interpretations in lucid terms: "Observation and experiment are not the bedrock upon which science is built; rather they are handmaidens to the rational activity of constituting knowledge claims through argument" (p. 555). Similarly, Lawson (2003) has emphasized that:

> A central activity for scientists is to construct and use arguments about which of the imaginative conjectures for a puzzling phenomenon are the most convincing in light of that evidence and, of course, to obtain additional evidence when the available evidence is insufficient or lacking. *(p. 1387)*

Osborne, Erduran, and Simon (2004), used the analytic framework developed by Toulmin (1958), for the design and evaluation of learning environments that support the teaching of argumentation in a scientific context. Toulmin's argu–

ment pattern was one of the first to challenge the "truth"-seekingrole of argument and to consider, instead, the rhetorical elements of argumentation and their function. For Toulmin, the essential elements of argument are claims, data, warrants, and backings. At the base of all arguments is a claim, essentially an unwarranted assertion that a proposer believes has the status of a universal truth. Arguments, however, normally rely on evidence or justifications that consist of data related to the claim by a warrant. Warrants, in turn, may be dependent on a set of underlying theoretical presumptions, which are often implicit. An important aspect of Toulmin's framework is that arguments are plausible inferences and not the deductive syllogisms, which "logical positivism" used to establish the objectivity of scientific explanations. In the first phase of the project, working with 12 junior high school science teachers from the greater London area, the main emphasis was to develop sets of materials and strategies to support argumentation in the classroom. Analysis of the data showed that there was significant development in the majority of teachers' use of argumentation across the year. In phase two of the project, teachers taught the experimental groups a minimum of nine lessons which involved socioscientific or scientific argumentation. In addition, these teachers taught similar lessons to a comparison group at the beginning and end of the year. Using a framework for evaluating the nature of the discourse and its quality developed from Toulmin's argument pattern, the findings show that there was improvement in the quality of students' argumentation. Research reported by these authors has been extended by Erduran, Simon, and Osborne (2004). Similarly, Simon, Erduran, and Osborne (2006) have shown that teachers' facilitation of argumentation has helped to guide the development of in-service materials and to identify the barriers to learning in the professional development of less experienced teachers.

According to von Aufschnaiter, Erduran, Osborne, and Simon (2008) there is an ongoing debate about whether poor performance in argumentation is a result of a lack of general competencies or whether it is caused by insufficient content-specific knowledge (p. 102). In order to pursue this and other issues, these authors investigated junior high school students' (in the greater London area) processes of argumentation and cognitive development in science and socioscientific lessons. Classroom discussions based on quality and frequency of students' argumentation were analyzed using a schema based on the work of Toulmin (1958). In parallel, students' development and use of scientific knowledge was investigated by drawing on a schema for determining the content and level of abstraction of students' meaning-making. One of the argument-basedlessons (out of a total of nine), included in the classroom was the following:

> When you exercise your skin gets redder especially on your face. Which of the following statements explains best this observation?: (a) Your blood pressure increases causing more blood to the surface of your skin; (b) Your blood is pumped to the surface for gaseous exchange to occur; (c) Your

> blood carries more oxygen and is therefore a deeper color; (d) Your blood gets closer to the surface for excess heat to be lost. *(p. 107)*

According to the authors, the structure of argument for this lesson involved: "Competing theories—ideas and evidence." Results obtained showed that: (a) When engaging in argumentation students draw on their prior experiences and knowledge; and (b) Such activity enables students to consolidate their existing knowledge and elaborate their science understanding at relatively high levels of abstraction. Finally, the authors concluded: "The major implication of this work for developing argumentation in the classroom is the need to consider the nature and extent of students' content-specificexperiences and knowledge prior to asking them to engage in argumentation" (p. 101). Based on Toulmin's (1958) framework, in order to measure the quality of argumentation, these authors looked for: claims, data, warrants, and rebuttals. According to Kuhn (1993), the ability to use rebuttals is the most complex skill and an individual must "integrate an original and alternative theory, arguing that the original theory is more correct" (p. 145). Von Aufschnaiter et al. (2008) concurred that the ability to construct a rebuttal or counterargument is crucial to the development of knowledge and the skill of argumentation.

Nussbaum, Sinatra, and Poliquin (2008) hypothesized that instruction in the criteria of scientific arguments, in combination with constructivist epistemic beliefs, would produce greater learning about physics concepts. The study was a randomized experiment, where college undergraduates discussed, in pairs over the Web, several physics problems related to gravity and air resistance. Prior to their discussions, one-halfof the dyads received information on the nature of scientific arguments. All students were classified epistemologically as relativists, multiplists, or evaluativists (for details on these classifications discussed above, see Florence & Yore, 2004; and Yore et al. 2004). Results showed that students in the treatment group incorporated more scientific criteria into their discussion notes and accordingly developed better arguments on several dimensions. In addition, significantly more participants in the treatment group adopted the correct answer to one of the problems. Outcomes also differed in relation to students' epistemic beliefs. Specifically, multiplists were less critical of inconsistencies and misconceptions, and interacted with their partners less than other belief groups, whereas evaluativists interacted more critically, bringing up different ideas from their partners.

A customized online discourse system that supports scientific argumentation has been developed by Clark and Sampson (2007). Students used an online interface to build principles to describe data they had collected. These principles constituted the seed comments for online discussions. The software sorted students into discussion groups with other students who have different principles so that each discussion group could consider and critique multiple perspectives. Finally, the study explored the efficacy of this personally seeded approach based on a coding scheme that analyzes argument structure from a Toulmin perspective.

In a study designed to foster argumentation, Chang and Chiu (2008) asked 70 undergraduate science and non-science majors in Taiwan to provide written arguments about four socioscientific issues. Results showed that: (a) Science majors' informal arguments were significantly better than those of non-science majors; (b) Science majors made significantly greater use of analogies, while non-science majors made significantly greater use of authority; and (c) Both groups had a harder time changing their arguments after participating in a group discussion. According to the authors, in the study of argumentation in science education, scholars have often used Toulmin's framework of data, warrant, backing, qualifiers, claims, and rebuttals. However, in their work, the authors found that Lakatos' framework is also a viable perspective, especially when warrant and backing are difficult to discern, and when students' arguments are resistant to change. This framework highlights how the "hard-core" of students' arguments about socio-scientific issues does indeed seem to be surrounded by a "protective belt" and thus difficult to alter. It is plausible to suggest that the resistance students offer in order to protect their epistemological and content-based beliefs is similar to that of the scientists (for details of various historical episodes, see Chapter 2).

Theoretical and empirical research on argument and argumentation in science education has intensified over the last two decades, which has been reviewed by Sampson and Clark (2008). The term "argument" in this review refers to the artifacts that a student or a group of students create when asked to articulate and justify claims or explanations, whereas the term "argumentation" refers to the process of constructing these artifacts. The intent of this review is to provide an overview of several analytic frameworks that science educators use to assess and characterize the nature of scientific arguments in terms of three focal issues: structure, justification, and content. To highlight the foci, affordances, and constraints of these different analytic methods, the review of each framework includes an analysis of a sample argument. This review suggested two overarching implications:

1. Much research to date has focused on very atomized aspects of students' arguments. While this atomized approach has proven fruitful, future research will hopefully include more holistic considerations of the quality of the arguments that students articulate. This work, however, will require new approaches that examine the structural, conceptual, epistemic, and social aspects of argument generation in a more synergistic fashion, rather than looking at each of these aspects independently.
2. Much research on argument in science education has thus far focused on the identification of patterns and themes in students' arguments (e.g., "students tend to produce arguments that lack sufficient justification" or "students tend to produce arguments that have a simplistic structure") rather than focusing on the underlying reasons for these patterns.

Argumentation has become an increasingly recognized focus for science instruction, as a learning process and as a window onto the epistemic work of science. However, according to Bricker and Bell (2008) only a small set of theoretical conceptualizations of argumentation have been deployed and investigated in science education, while a plethora of conceptualizations have been developed in the interdisciplinary fields associated with science studies and the learning sciences.

Research reviewed in this section shows that the use of argumentation in the classroom can be a useful teaching strategy in order to enhance students' conceptual understanding. Despite the promise and the fruitfulness of this approach, some difficulties were also encountered:

a. Argumentation is basically a dialogic event. Given the authoritarian classroom environments in most parts of the world, opportunities for dialogic discourse are limited.
b. We need to develop students' content-specific knowledge before asking them to engage in argumentation. In other words in order to arouse students' interest and curiosity it is helpful if the arguments are based around a particular topic of the science curriculum.
c. The ability to construct a rebuttal or counter-argumentis crucial to develop the skill of argumentation.
d. Most of the studies have used Toulmin's framework. However, it appears that a Lakatosian framework is also viable, especially when warrant and backing are difficult to differentiate and students' arguments are resistant to change.
e. Instead of focusing on the identification of patterns in students' arguments, there is a need to research the underlying reasons for these patterns.

At this stage it is important to note that argumentation is one of the most important topics of research in the three major science education journals (*International Journal of Science Education*, *Journal of Research in Science Teaching*, *Science Education*), in the period 2003–2007 (cf. Lee, Wu, & Tsai, 2009). Furthermore, the following articles based on argumentation were included among the 10 most cited in these journals: Osborne et al. (2004), Simon et al. (2006), and Erduran et al. (2004).

Explicit and Reflective Versus Implicit Inquiry-Oriented Instruction (n = 11)

Although the *National Science Education Standards* (NRC, 1996) recommend that elementary teachers help their students understand science content and science as a discipline through inquiry teaching, it has been shown that teachers themselves do not generally have informed conceptions of the NOS (Lederman, 1992). Portraying science in such a way that goes beyond thinking of science as a body of knowledge requires an understanding of the NOS (Hodson, 2009). Simply engaging implicitly in scientific inquiry activities is not sufficient in helping teachers obtain better understandings of NOS content. It is plausible to suggest that teachers can develop informed conceptions of the NOS through an explicit reflective approach.

Akerson, Abd-El-Khalick, and Lederman (2000) used a reflective activity-based approach in a science-methods course, to facilitate elementary teachers' conceptions of the NOS (especially, empirical, tentative, theory-laden, imaginative, creative, social, and cultural aspects). Given the participating teachers' limited science content background, activities selected for the study were generic in nature and not content specific. Results indicated that participating teachers made substantial gains in their understanding of target aspects of NOS. Based on their experience these authors finally raised a very important issue: "it cannot be overemphasized that we believe that NOS instruction is best undertaken in the context of science content courses. However, NOS instruction is almost nonexistent in the agendas of the mostly traditional science content courses offered in disciplinary departments" (p. 313). This clearly shows the need for introducing the NOS (and also the HPS) not only through explicit activity-based approaches but also through the inclusion of appropriate material in science content courses (biology, chemistry, physics, etc.). Such teaching approaches (HPS- and NOS-enriched) have the additional advantage of providing students the understanding that issues related to scientific practice (alternative interpretations of data, conflicts, controversies, theory-laden observations) are an integral part of science. Implementation of such approaches necessarily requires the writing of appropriate textbooks and other materials. Consequently, evaluation of science textbooks based on the NOS- and HPS-based criteria is an important area of research.

Akerson and Volrich (2006) have focused on a pre-service teacher's efforts to explicitly emphasize the NOS elements in her 1st grade internship classroom. The study assessed the change in 1st grade students' views of the inferential, tentative, and creative NOS as a result of the explicit instruction. Morgan (the teacher) held appropriate views of the NOS, had the intention and motivation to teach the NOS, and had a supporting experience explicitly emphasizing the NOS embedded in physics content to peer college students. Data sources included weekly classroom observations of explicit NOS science lessons taught by Morgan, interviews with Morgan to determine that her views of the NOS were informed and that she would have the NOS content knowledge to teach in line with recommended reforms, and interviews of the 1st grade students pre- and post-instruction to determine the influence of Morgan's instruction on their views of observation and inference, the tentative NOS, and the creative and imaginative NOS. Data were analyzed to determine: (a) the approaches Morgan used to emphasize NOS in her instruction; and (b) students' views of NOS pre- and post-instruction to track change in their views. It was found that Morgan was able to explicitly emphasize the NOS using three teacher-designed methods, and that the influence on student views of the inferential, tentative, and creative NOS was positive. At this stage readers would like to know how Morgan used explicit NOS instruction through classroom interactions with her 1st grade students. The following account is based on the section "Explicit introduction to NOS aspects" (p. 386). Morgan wanted to emphasize observation and inference as one of the NOS aspects in her instruction. Three days after her introductory discussion on students' ideas of science, she conducted an explicit lesson on observations and inferences. In this lesson she first asked students whether they knew what an observation was. Students generally understood that an observation was something you "see." Morgan asked them to extend this view to things you can smell, touch, hear, and taste. She reminded them of earlier studies on the five senses, and helped them to realize that the use of the five senses would constitute observations. She then had them observe a stuffed animal and asked them to smell it, see it, touch it, and listen to it. She asked them to tell her what they thought it was. When the students told her it was a toy wolf, she said, "You are inferring it is a toy wolf from your observations. That is what scientists do—they use their observations to make inferences about things—it helps them understand what they are doing" (p. 386).

A professional development program for teaching the NOS to elementary teachers has been designed by Akerson and Hanuscin (2007). The program consisted of 84 hours of monthly workshop time over a period of three years, and was implemented at a university town in Midwest USA. Workshop activities were selected to be generic in nature (not content specific) given the participants' limited science content backgrounds. During the workshops the teachers wrestled with new ideas, as can be observed from the following exchange with respect to the "tubes" activity:

Facilitator: Can you know what is in the tube? And does it matter? Irrespective of how you built your tube, your tube still works.

Carrie: Well it is interesting because we can't physically observe everything, but it is a study of patterns, and we figure out what those patterns mean. There is something about absolute truth. We will never know for sure.

Facilitator: The interesting thing is we can kind of know what happens even though we can't see it. Most of the time in science all we can do is make observations, and try to construct a model of reality to draw inferences that behaves, and predicts what happens in reality. It is not necessary to know how the tube works in the inside. Why don't we say there is a little guy [or girl] pulling on the strings inside the tube?

Charlotte: Because it is not reasonable.

Facilitator: But is it more reasonable that we say the whole universe was the size of a needle, became very dense and then exploded into what it is now? Big Bang Theory. How reasonable is that? . . . Does this thing with absolute truth have anything to do with teaching science in the classroom?

Charlotte: It is so much easier that you teach them that it is the truth. You don't have them question how we know this.

Facilitator: That is a valid point. But when you are doing this, what part of science are you doing?

Charlotte: Body of knowledge. You're not teaching them about the processes, or how science works at all. (reproduced in Akerson & Hanuscin, 2007, pp. 659–660)

This clearly shows how in a explicit reflective approach teachers' understanding of the NOS can be enhanced. Interestingly, the authors also reported that after the workshops the teachers requested a curriculum that provided NOS activities within inquiry lessons. However, they soon realized that changing the curriculum cannot be accomplished readily and so they spent considerable time adapting their own curriculum to meet these instructional goals. This experience has important implications for the development of science curriculum. Based on research in science education (this chapter can be particularly helpful), teachers can take the initiative to change their classroom environment by going beyond the existing curriculum.

Khishfe and Lederman (2006) investigated the influence of two different explicit instructional approaches in promoting more informed understandings of the NOS among 9th grade students. Participants (n = 42) comprised two groups of intact sections. Participants in the two groups were taught environmental science by their regular classroom teacher, with the difference being the context in which the NOS was explicitly taught. For the "integrated" group, the NOS instruction was related to the science content about global warming. For the "nonintegrated" group, the NOS was taught through a set of activities that specifically addressed NOS issues and were dispersed across the content about global warming. The treatment for both groups spanned six weeks and addressed a unit about global warming and the NOS. An open-endedquestionnaire, in conjunction with semi-structuredinterviews, was used to assess students' views before and after instruction. Results showed improvements in participants' views of the NOS regardless of whether the NOS was integrated within the regular content about global warming. Comparison of differences between the two groups showed slightly greater improvement in the informed views of the integrated group participants. Overall results did not provide any conclusive evidence in favor of one approach over the other.

The relationship between instructional context (integrated and non-integrated) that explicitly teaches about the NOS and students' views of the NOS across different disciplines, has been explored by Khishfe and Lederman (2007). Participants were three teachers and their students, which comprised six classes of 89 9th-gradersand 40 10th/11th graders. Each teacher taught two intact sections of the same grade level within a specific science discipline. The treatment for all groups involved teaching a 5–6 weekunit that included the science content and the NOS. The two intact groups learned about the same content, the only difference was the context of the NOS instruction (integrated or non-integrated).An open-endedquestionnaire, followed by interviews, was used to assess change in participants' views. Results showed improvement in students' views of the NOS regardless of whether the NOS instruction was embedded within the context. Authors concluded that it was not possible to make claims about whether one instructional context is more effective than another (similar results were obtained in Khishfe & Lederman, 2006).

Khishfe (2008) has investigated the development in students' views of the NOS in the context of an explicit inquiry-orientedinstructional approach. Participants were 18 7th-gradestudents who were taught by a teacher with "appropriate" knowledge about the NOS. The intervention spanned about three months. During this time, students were engaged in three inquiry-orientedactivities that were followed by reflective discussions of the NOS. The study emphasized the tentative, empirical, inferential, and creative aspects of the NOS. An open-ended questionnaire, in conjunction with semi-structuredinterviews, was used to assess students' views before, during, and after the intervention. Before instruction, the majority of students held naive views of the four NOS aspects. During instruction, the students acquired more informed and "intermediary" views of the NOS aspects. By the end of the intervention, the students' views of the NOS aspects had developed further still into informed and "intermediary."

Abd-El-Khalickand Akerson (2004) identified factors (*learning ecologies*) that mediated the effectiveness of an explicit reflective instructional approach for learning as conceptual change on pre-serviceelementary teachers' views of the NOS. An examination of the development of the focus group participants' NOS views indicated that the effectiveness of the intervention was mediated by motivational, cognitive, and worldview factors. These were related to internalizing the importance and utility of teaching and learning about the NOS, exhibiting a deep processing approach to learning, and viewing science and religion as two distinct rather than opposing enterprises. The construct of "learning ecologies" has been the subject of criticism and dissatisfaction with respect to its "fruitfulness" as part of a theory in science education research (Johnston, Southerland, & Sowell, 2006). According to this criticism, *learning ecology* adds little to the construct of *conceptual ecology*, used frequently in the widely used conceptual change theory (CCT) in science education (Posner et al., 1982). In their response, Abd-El-Khalickand Akerson (2007) acknowledged that they were not trying to develop a new theory or extend CCT. Furthermore, *learning ecology* and *conceptual ecology* are not equivalent entities. According to these authors, "learning" in "learning ecology" was meant to focus attention on the hypothesis that learning is not merely cognitive and rational, but also social and collaborative (p. 191). Again, the revisionist CCT (Strike & Posner, 1992), did recognize the affective, motivational, and other extrarational factors, but still maintained its central commitment, namely, that learning is an individual, cognitive, and rational activity.

Significantly, Abd-El-Khalickand Akerson (2007) pointed out that the difficulty with CCT is more profound and perhaps can be traced to the fact that it is heavily imbued with Kuhn's (1970) scientific revolutions. In accepting Kuhn's ideas, CCT almost completely ignored the very process by which learners abandon a previously held paradigm in favor of an alternative paradigm. Within a Lakatosian framework (Lakatos, 1970), it is plausible to suggest that both learning and conceptual ecology, represent an auxiliary hypothesis or a background assumption, and would form part of the theoretical system when CCT is empirically tested. Within this perspective the authors suggested that in the absence of a robust alternative, they are not prepared to abandon CCT, as it still remains a useful theory. Finally, in their response, Sowell, Johnston, and Southerland (2007) concluded: "We see the reasoning behind their [Abd-El-Khalick& Akerson, 2007] term 'learning ecology' as twofold: emphasizing the social and collaborative aspects of conceptual change, while simultaneously critiquing the overly individualistic nature of conceptual change theory itself" (p. 195).

Teacher education programs have met with limited success in improving teachers' understanding of the NOS. Hanuscin, Akerson, and Phillipson-Mower (2006) designed explicit NOS instruction in a physical science content course for pre-serviceelementary teachers. Although initially teachers held a number of views inconsistent with science education reforms, four modes of explicit-and-reflective interventions resulted in favorable changes.

Akerson, Buzzelli, and Donnelly (2008) tracked the influence of explicit reflective instructional methods on cultural values, ethical and intellectual development, and the relationship of these with pre-service teachers' views of the NOS. The researchers used the Views of Nature of Science Form B (VNOS B) to describe NOS views, the Learning Context Questionnaire (LCQ) to classify pre-serviceteachers' ethical and intellectual positions using Perry's scheme, and the Schwartz Values Inventory (SVI) to measure pre-service teachers' cultural values. Perry's (1999) scheme explored adult cognitive development and related it to ways of learning. Authors used this scheme to understand students' inability to retain new ideas. Perry's scheme lists nine positions of cognitive development for adult learners. Perry, however, made no presumptions about the length of time a person may remain in a position and thus the positions are more fluid, unlike Piaget's developmental stages. The interventions took place in two concurrent courses: a science-methodscourse, and a foundation of early childhood course. The science-methods course explicitly emphasized the NOS throughout the semester, and the foundations of early childhood course reinforced these ideas through cultural activities that stressed empirical evidence. Analysis of data showed relationships between pre-serviceteachers' Perry positions and responses on the VNOS-B,with those at higher positions exhibiting more informed views of the NOS. Relationships between pre-serviceteachers' NOS views and their cultural values were identified, such as those at the dualism position holding achievement more highly for scientists than those at other Perry positions. The values pre-serviceteachers held personally were different from those they held as important for scientists.

Research reviewed in this sections reveals that: (a) The NOS instruction is best undertaken in the context of science-context courses (biology, physics, chemistry, etc.); (b) Issues related to scientific practice (alternative interpretations of data, conflicts, controversies, theory-ladenobservations) are an integral part of science and as such can facilitate a better understanding of the NOS; (c) Introducing the NOS in the classroom may require a change in the curriculum, which is not accomplished readily; (d) There is a need for explicit and reflective teaching strategies in order to introduce the NOS in the classroom; (e) There is a need for going beyond conceptual ecology (based on Posner et al., 1982) to learning ecology that emphasizes the social and collaborative aspects of learning besides the cognitive and rational (cf. Strike & Posner, 1992). In contrast, Clough (2006), however, has emphasized the importance of misconceptions about the NOS which are reinforced by out-of-schoolportrayals of science observed on television and other media. Consequently, besides explicit and reflective teaching we need to take account of how students learn and learners' responses to the demands of conceptual change.

Use of NOS-enriched Instructional Materials (n = 7)

Brickhouse, Dagher, Letts, and Shipman (2000) have shown how students may develop their ideas about theory, evidence, and the science/religion interface during a one-semester,non-majorcollege astronomy course. The course is based on a textbook (Shipman, 1980) and dealt with the following questions: How do we find black holes? What is a black hole? How did the universe begin? How did life originate, and become more complex? These are interesting and intriguing issues, and a major objective of the course was to show how scientists probe such questions and that scientific theories are not mere hunches but, rather, have undergone considerable testing and debate. For example, the first chapter of the course textbook, on black holes, draws heavily from Kuhn (1962) and emphasizes that scientific knowledge is subject to change in complex ways. This course and the material (Shipman, 1980) is a good example of an NOS-enrichedcontent-based course. As a follow-up,Dagher, Brickhouse, Shipman, and Letts (2004) explored college students' representations about the nature of theories during their enrollment in a large astronomy course with instruction designed to address NOS issues. Students' notions of theory were classified under four main categories: hypothesis, idea with evidence, explanation, and explanation based on evidence. Students' conditions for deciding whether a given idea is a scientific theory or not were classified under six criteria: content domain, convention, evidence, mathematical content, methodology, and tentativeness. With respect to theories and laws the authors concluded, "students are likely to implicitly rank them on a misleading hierarchy of certainty or a naïve sense of tentativeness. They will continue to believe that hypotheses become theories and theories become laws" (p. 754). Authors underscored the need to explicitly address the nature of proof in science and issues of tentativeness and certainty students associate with scientific theories, and provide students with more opportunities to utilize the language of science.

Difficulties encountered by a graduate teaching assistant as she taught an inquiry-based physics course for elementary education, have been reported by Volkmann and Zgagacz (2004). At the beginning of the course the teacher believed that the best and most efficient way to teach students the facts was to tell them, which was quite consistent with Schwab's (1962) description of science teaching: "Science is taught as a nearly unmitigated *rhetoric of conclusions* in which the current and temporary constructions of scientific knowledge are conveyed as empirical, literal and irrevocable truths" (p. 24). This teacher's realist understanding of physics was at odds with the ways in which the physics community constructed meaning within working groups, collaborations, and exchange of ideas at conferences. Furthermore, she ignored the guidelines for inquiry teaching suggested by *Powerful Ideas* (American Association of Physics Teachers, AAPT, 1996). Authors concluded that fundamental beliefs about the NOS support certain orientations, and if those beliefs remain unchallenged, then the orientation is unlikely to change.

In a Delphi style study, Osborne, Collins, Ratcliffe, Millar, and Duschl (2003) consulted 23 experts engaged in the study of science and its communication, based on the following groups: leading scientists, historians, philosophers and sociologists of science, science educators, science teachers, and scholars working on the public understanding of science. These experts recommended that the following "ideas-about-science"could be taught in school science: (1) Scientific Method and Critical Testing; (2) Creativity; (3) Historical Development of Scientific Knowledge; (4) Science and Questioning; (5) Diversity of Scientific Thinking; (6) Analysis and Interpretation of Data; (7) Science and Certainty; (8) Hypothesis and Prediction; and (9) Cooperation and Collaboration in the Development of Scientific Knowledge. Interestingly, the experts assigned the highest priority to the teaching of "Scientific Method and Critical Testing," with reasons such as: (a) "core process on which the whole edifice of science is built"; (b) "central thrust of scientific research"; and (c) "careful experimentation is used to test hypotheses." Martín-Díaz(2006) has reported that teachers in Spain would agree with various aspects of "ideas-about-science"(Osborne et al., 2003), such as: provisional, experimental, and predictive nature of scientific knowledge. As a follow-up, Bartholomew, Osborne, and Ratcliffe (2004) designed a study to teach students "ideas-about-science."A group of 11 U.K. teachers participated in the study over a period of a year, to teach aspects of the NOS, its process, and its practices. The teachers, who taught science in a mix of elementary, junior high, and high schools, were asked to teach a set of "ideas-about-science"for which consensual support had been established using a Delphi study in the first phase of the project (Osborne et al., 2003). Data were collected through field notes, videos of the teachers' lessons, teachers' reflective diaries, and instruments that measured their understanding of the NOS and their views on the role and value of discussion in the classroom. Based on the data, authors concluded that there are five critical dimensions that distinguish and determine a teacher's ability to teach effectively about science: (1) Teachers' knowledge and understanding of the nature of science; (2) Teacher's conceptions of their own role; (3) Teachers' use of discourse; (4) Teachers' conception of learning goals; and (5) The nature of classroom activities. Furthermore, authors suggested that these dimensions are an important means of identifying salient aspects of pedagogy for initial and in-servicetraining of science teachers for curricula that incorporate elements of "ideas-about-science."

Ford and Wargo (2007) have characterized engagement in scientific practice in terms of "3Rs"—routines, roles, and responsibilities—and argued that there is a misalignment between the 3Rs of scientific practice and the practices common in classrooms, and that this misalignment poses a considerable obstacle for beginning teachers who attempt to implement reform pedagogy. As part of a secondary methods course, 16 pre-serviceteachers (PSTs) participated in two exemplar activities designed to engage them in scientific practices. The PST performances suggest that, at least initially, they did not consider authentic scientific practices appropriate for classroom activities, implying a pedagogical repertoire dominated by the 3Rs of traditional classrooms. PST performances, however, evidenced a shift in the 3Rs from those common in classrooms to those required by these activities, suggesting that their visions for classrooms are malleable and underlining the importance of aligning the 3Rs of scientific and classroom practices during teacher preparation.

Niaz (2008b) has designed a study to facilitate in-servicechemistry teachers' understanding of the NOS and what "ideas-about-science"can be included in the classroom. The study is based on 17 in-serviceteachers who had registered for an 11-weekcourse on "Epistemology of Science Teaching" as part of their Master's degree program. The course is based on 17 readings drawing on the NOS and its critical evaluation. Among others, the following readings formed an important part of the course: Blanco & Niaz, 1997; Campanario, 1999; Lederman et al., 2002; Lin & Chen, 2002; Matthews, 2004; Mellado, 2003; Niaz, 2001a; Osborne et al., 2003; Scharmann & Smith, 2001; Smith & Scharmann, 1999. Course activities included written reports, classroom discussions based on participants' presentations, and written exams. Based on the results obtained this study facilitated teachers' understanding of the following aspects of the NOS:

a. Experimental data need to be interpreted carefully due to underdetermination of theories by data.
b. Kuhn's normal science manifests itself in the science curriculum through the scientific method and wields considerable influence.

c. The trilemma posed by Collins (2000), that is, creation of new knowledge ↔ Kuhn's normal science ↔ teaching the NOS, provided a big challenge and was thought provoking.
d. Of the different aspects of the NOS suggested by experts, these teachers endorsed the following as most important: Creativity, Historical development of scientific knowledge, Diversity of scientific thinking, and Scientific Method and Critical Testing.
e. With respect to the contradiction between the positions of Lederman et al. (2002) and Osborne et al. (2003), few supported the position of the latter—inclusion of scientific method in the classroom—and a majority supported that of the former—scientific method as a myth.
f. Participants were critical of the present stage of research with respect to the scientific method and suggested the introduction of history, philosophy, and epistemology of science to counteract its influence.

Wong, Hodson, Kwan, and Wai Jung (2008) have shown how a crisis can be turned into an opportunity for designing NOS-enrichedmaterials. These authors interviewed key scientists involved in research on severe acute respiratory syndrome (SARS), together with analysis of media reports and documentaries produced during and after the SARS epidemic, which revealed many interesting aspects of the NOS. This novel insight into practice in the rapidly growing field of molecular biology was used in the development of instructional materials for use in the pre-serviceand in-serviceteacher education program at the University of Hong Kong. Elements of the NOS and scientific inquiry identified in the scientific research on SARS were explicitly emphasized in the instructional materials. The contemporary real-lifecontext of SARS was found to be effective in promoting student teachers' understanding of the NOS, in terms of: (a) Realization of inseparable links between science and the social, cultural and political environment; (b) Deeper understanding of how science and technology impact on each other; and (c) Richer appreciation of the processes of authentic scientific inquiry and the humanistic character of scientists.

Conceptualization of the NOS and how students interpret and evaluate conflicting evidence regarding a socioscientific issue, has been investigated by Sadler, Chambers, and Zeidler (2004). Eighty-fourhigh school students participated in the study by reading contradictory reports about the status of global warming and responding to questions designed to elicit ideas pertinent to the research goals. Thirty students were interviewed in order to triangulate data from the written responses. Data were analyzed using a qualitative methodological approach. Participants displayed a range of views on three distinct aspects of the NOS: empiricism, tentativeness, and social embeddedness. Findings indicate that interpretation and evaluation of conflicting evidence in a socioscientific context is influenced by a variety of factors related to the NOS, such as data interpretation and social interactions, including individuals' own articulation of personal beliefs and scientific knowledge.

Research reviewed in this section reveals the need for NOS–enriched content–based courses. Despite efforts to change, students continue to believe that hypotheses become theories and theories become laws. In order to change students' understanding of the NOS, it was found that teachers' knowledge of the NOS and the nature of classroom activities, play an important role. It seems that a crisis can be turned into an opportunity for designing NOS–enrichedinstructional materials. An example of such content–based materials is provided by recent research on SARS, which facilitates an understanding of the links between science and the sociocultural environment.

Use of History-based Instructional Materials (n = 3)

Pocoví (2007) studied the effect of history–basedinstructional material designed to foster the analysis of the ontology of field lines to facilitate freshmen students' conceptual understanding at a university in Argentina. The historical evolution of this concept shows that early scientists such as Faraday assigned material properties to these lines on some occasions. Thus, some of Faraday's writings provide examples where the ontology that he assigned to the lines of force is evidenced through the predicates he used with them. These examples were used in the present study to teach the nature of lines of force through the analysis of the predicates that can be used with the geometric conception of lines and those that correspond to a matter–basedview of them. Results obtained showed that the experimental group (using the historically rich instructional material) performed significantly better than the control group.

Development, use, and analysis of an educational tool inspired by the history of the optical mechanism of vision has been reported by Hosson and Kaminski (2007). These authors investigated 12–year–oldFrench students' reasoning about vision as something coming either from the object or from the eye. Some students thought that light penetrated the eye only when they were dazzled. Such ideas can be found in the ancient and medieval history of science. Ancients disagreed about the direction of vision until Alhazen opened the way to a consensus, arguing in the eleventh century that light could be a stimulus for the eye. This study designed a short drama entitled, "Dialogue on the ways that vision operates," based on historical elements, especially the controversy over the direction of vision and Alhazen's ideas about light. This text was integrated in a teaching–learningsequence and experimented with six pairs of students. Findings showed that the learning process can take advantage of the opportunity offered to the students to identify themselves with the scientists portrayed in the drama.

Abd–El–Khalick(2005) has investigated the influence of a philosophy of science (POS) course on pre–servicescience teachers' views of the NOS, perceptions of teaching about the NOS, and instructional planning related to the NOS. Participants were 56 undergraduate and graduate students enrolled in a two sci–

ence–methodscourse sequence, in which they received explicit, reflective NOS instruction. Ten of these participants were also enrolled in a subsequent graduate survey POS course. The *Views of Nature of Science Questionnaire—Form C*, coupled with individual interviews was used to assess participants' views of the NOS at the beginning and conclusion of the study. Students enrolled in the science methods courses, wrote the following NOS–specificreflection papers: (a) Participants were asked to discuss the NOS ideas presented in McComas (1996) and compare them with their own views; and (b) Participants read the prologue for Penrose (1994) and answered the following question: "Do the ideas in this reading fit our discussions of some aspects of NOS?" Participants enrolled in the POS course wrote four reflection papers based on several original works of twentieth–century philosophers of science. Some of the readings were on the following: The dinosaur extinction controversy (Alvarez & Azaro, 1990); the Copernican revolution (Kuhn, 1985); the Michelson–Morleyexperiment (Lakatos, 1970); Cold fusion and verification of the relativity theory (Collins & Pinch, 1993). The complete list is provided by the author in Table 2 (Abd–El–Khalick,2005, p. 21) which provides a flavor of how science progresses through rival interpretations of data and the ensuing controversies. Results indicated that, compared with participants enrolled in the methods courses, POS course participants developed deeper, more coherent understandings of the NOS. These are significant findings and can provide insight for future research. However, the author of the study interprets the findings with some caution, as: (a) The number of participants in the POS group was relatively small; (b) The researcher was the instructor for the two methods and POS course; and (c) The POS course was not representative of "typical" POS courses.

Research reviewed in this section shows that explicit, history–basedinstructional materials can help to enhance students' understanding of the NOS. However, it seems that such experiences are perhaps more suitable after students have gained some preliminary knowledge of the NOS. For example, in one course students were provided readings from twentieth–centuryphilosophers of science based on the following controversies. It seems that the transition from the NOS ideas in general to understanding actual historical episodes is a fruitful strategy.

Use of Technology-based Historical Materials (n = 2)

A framework for scaffolding epistemic aspects of inquiry that can help students understand inquiry processes in relation to the kinds of knowledge (NOS), has been presented by Sandoval and Reiser (2004). The framework is based on a technology–supportedinquiry curriculum for evolution and natural selection that focuses students on constructing and evaluating scientific explanations for natural phenomena. The design has been refined through cycles of implementation, analysis, and revision that have documented the epistemic practices students engage in during inquiry, indicate ways in which designed tools support students' work, and suggest necessary additional social scaffolds. Findings suggest that epistemic tools can play a unique role in supporting students' inquiry, and provide a fruitful means for studying students' scientific epistemologies.

Crawford et al. (2005) used a technology-enhanced inquiry approach (*The Galapagos Finches* software) to facilitate prospective teachers' understanding of evolution. This study also revealed that even teachers with research experience failed to understand the diverse methods used by scientists.

Use of Science Apprenticeship Programs (n = 2)

Reform efforts emphasize teaching science to promote contemporary views of the NOS and scientific inquiry. It is plausible to suggest that inquiry activities similar to those of scientists provide a learning context conducive to developing knowledge about the methods and activities through which science progresses, and consequently developing desired views of the NOS. However intuitive this may appear, more work needs to be done to support the claim that engaging in scientific inquiry alone enhances conceptions of the NOS.

Based on their experience of teaching the NOS through inquiry to prospective elementary teachers, Bianchini and Colburn (2000) emphasized the teacher's pivotal role in initiating discussions of what science is and how scientists work, and the strengths and limitations of using classroom-basedresearch to investigate NOS instruction.

Ryder, Leach, and Driver (1999) reported a study in which final year undergraduate students in the UK had the opportunity to work as research apprentices for up to eight months. During this period of socialization into the practice of scientific research in universities, students were exposed to scientific discourse by scientists who were currently engaged in conflicts of ideas which are yet to be resolved. Many of the students showed significant development in their understanding of how lines of scientific enquiry are influenced by theoretical developments within a discipline. Despite improvement, a majority of the students indicated that knowledge claims could be proved absolutely and that empirical data was the sole ground for proof. According to the authors, this finding supports that of Rowell and Cawthron (1982), who reported that university students tend to advance a Popperian view in which scientific ideas are either accepted or falsified on the basis of empirical data. This, contrasts sharply with the ideas of Kuhn (1962), Lakatos (1970), and Giere (1999, 2006), according to which empirical data do not unambiguously lead to scientific theories. Furthermore, according to Ryder et al. (1999) students' images of science involve tacit knowledge (Polanyi, 1967) which is rarely made explicit but which nevertheless informs action. Similarly, Mortimer (1995) has argued that a student exhibiting a particular image of science in one context may deploy a different image in another, and this range of images of science constitutes a profile of understandings in specific concept areas.

Schwartz, Lederman, and Crawford (2004) studied improvements in understanding of the NOS during a science research internship course for pre-service secondary science teachers. In addition to the research component, the course included seminars and journal assignments. Interns' views of the NOS were assessed in a pre/post format using the Views of Nature of Science questionnaire [VNOS-C],and interviews. Results indicated substantial gains in NOS knowledge, due to the following factors: (1) Reflection; (2) Context; and (3) Perspective. Reflective journal writing and seminars had the greatest impact on NOS views. The science research component provided a context for reflection. The interns' role perspective appeared to impact their abilities to effectively reflect. Interns who assumed a reflective stance were more successful in deepening their NOS conceptions. In contrast, those who maintained a scientists' identity were less successful in advancing their NOS views through reflection. These results suggest that simply observing scientists in an apprenticeship program might not be very helpful, unless the participants have the appropriate framework for understanding and reflecting on what they observe.

Efforts to promote more realistic conceptions about science are often limited by teachers' inexperience of real science. Bencze and Elshof (2004) have examined an "inductive-deductive, dialectic immersion" approach towards assisting teachers in developing more realistic conceptions about science, along with corresponding revised perspectives about science teaching. Three secondary teachers of science with minimal science research experience engaged in a case study of science in action, specifically in an episode of northern alpine ecological field research. Qualitative data analyzed by constant comparative methods suggested that these teachers shifted along a modernist through postmodern continuum. Authors concluded that science teachers can develop postmodern views about science and science teaching if given opportunities to induce and deduce propositions about real science.

Conclusions and Educational Implications

This chapter reviewed 94 articles published over a period of five years (2004–2008) in the following journals: *International Journal of Science Education* (n = 34), *Journal of Research in Science Teaching* (n = 28), and *Science Education* (n = 32). Distribution with respect to the country (where the study was conducted, in the case of more than one author), was the following: USA = 51, Taiwan = 9, UK = 7, Turkey = 5, Canada = 4, Australia = 2, France = 2, Israel = 2, Korea = 2, Lebanon = 2, New Zealand = 2, Argentina = 1, Estonia = 1, Hong Kong = 1, the Netherlands = 1, Sweden = 1, Zimbabwe = 1. This clearly shows that a major part of the work related to the HPS and the NOS was published in the USA, with smaller contributions from various other countries. All these studies emphasized some aspect of the HPS and its implications for science education. Furthermore, articles based on textbooks with an HPS perspective were not included in this study. In a previous study based on these three journals over a five–yearperiod (1998–2002), Tsai and Wen (2005) found 68 articles related to *History, Philosophy, Epistemology and Nature of Science* (out of a total of 802 articles, 8.5%). In a follow–up study, Lee, Wu, and Tsai (2009) found 71 articles related to this topic in the same three journals in the period 2003–2007. This shows the continued interest in this area of research. Some of the salient aspects of research reviewed in this chapter are the following:

a. It is interesting to study how research in NOS–relatedareas has evolved (longitudinal aspect) as compared to earlier studies. However, the following aspect is noteworthy: In his review, Lederman (1992) had noted, "both groups [students and teachers] have been berated for not understanding that scientific knowledge is necessarily tentative" (p. 352). Research reviewed in this study shows that despite some improvement, the tentative nature of scientific knowledge continues to be one of the most difficult aspects of the NOS in most parts of the world.
b. As compared to previous research, this study reveals that instead of evaluating the NOS views there is more concern for enhancing students' and teachers' understanding of the NOS and HPS. Furthermore, there seems to be some consensus that instead of implicit we need explicit and reflective teaching strategies embedded in the HPS–and NOS–orientedaspects of science content courses.
c. The use of argumentation in the classroom is one of the most novel aspects of research reviewed in this study. This research has centered on the historical context (e.g., controversies with respect to atomic models) and everyday life context (e.g., why does your skin get redder when you exercise?), and analyzed within the philosophical frameworks of Toulmin and Lakatos.

Slater (2008) has raised a provocative question for science education: How can we justify teaching false science? This, in turn, is based on the premise that we teach false science (e.g., Newtonian mechanics, Bohr's model of the atom). As a possible solution to the dilemma, Slater suggests, "the best way of teaching false science is by teaching it *as false*, but *illustratively*—incorporating a critical historical perspective into the science curriculum" (p. 541). This clearly shows the need for incorporating an HPS perspective in the science curriculum, in order to facilitate a better understanding of the dynamics of scientific progress. Similarly, Collins (2000) has endorsed the view that the introduction of an HPS perspective can help to counteract the "falsified history" found in most science textbooks.

Bartholomew et al. (2004) have emphasized the need to incorporate the following dimensions in order to facilitate students' and teachers' understanding of the NOS, within an HPS perspective: (1) Teachers' knowledge and understanding of the NOS; (2) Teachers' conceptions of their own role; (3) Teachers' use of discourse; (4) Teachers' conceptions of learning goals; and (5) The nature of classroom activities. According to the authors, of these five dimensions, only the first has been given sufficient attention in the literature. Furthermore, teachers' understanding of the various aspects of the HPS is essential for them to incorporate such a perspective in the classroom. It is important to emphasize that even if we have appropriately trained science teachers and an HPS-basedcurriculum—this by itself does not translate into improved practice and effective pedagogy. More research needs to be done in designing classroom studies based on an HPS perspective (cf. Niaz, 2011). Abd-El-Khalickand Lederman (2000) drew a similar conclusion in an extensive review of many studies based on teaching the NOS. Furthermore, this review also reveals that besides argumentation and reasoning, teachers' views of the NOS are also guided by intuition, emotion, prior epistemological beliefs, and extrarational factors (Sadler, 2004; Southerland et al., 2006).

Both Schwartz and Lederman (2008) and Yore et al. (2004) agreed that most scientists' epistemological views in their studies could be considered to be "evaluativist" (science knowledge is a temporary explanation that best fits the existing evidence and current thinking). Nonetheless, they have also called for a word of caution as engaging in authentic scientific inquiry as a successful member of the scientific community is not *necessarily* sufficient, in and of itself, to ensure informed conceptions of the NOS (Schwartz & Lederman, 2008, p. 763). As an example, it is interesting to observe a tension between the objective and subjective aspects of science, in the understanding of these scientists:

> researchers interviewed did not freely mention the creative human and social components of constructing knowledge claims in science and technology. It appeared as if these academics believed that they needed to constantly stress the objective science traditions of observations, measurements, accepted procedures, and canonical knowledge, and to avoid mentioning the subjective human dimension of making sense of these data in science and technology. *(Yore et al., 2004, p. 363)*

This provides important insight with respect to 'science in the making.' Do scientists generally report in their professional journals how and why they did an experiment? (Chapter 1 referred to this as an important aspect of the NOS.) This is a crucial issue and in the long run determines how textbooks are written and science is presented in the classroom (cf. Niaz & Maza, 2011, for the presentation of the NOS in general chemistry textbooks). A possible response to this dilemma is provided by Peter Medawar (Nobel laureate in medicine, 1960):

> It is no use looking to scientific "papers," for they not merely conceal but actively misrepresent the reasoning that goes into the work they describe. If scientific papers are to be accepted for publication, they must be written in the inductive style. The spirit of John Stuart Mill glares out of the

> eyes of every editor of a Learned Journal . . . Only unstudied evidence will do—and that means listening at a keyhole. *(Medawar, 1967, p. 151)*

This clearly has implications for science education if we want to understand how science is practiced by scientists. Listening at a keyhole could help in the resolution of the dilemma and would require collaboration among historians, philosophers of science, and science educators (cf. Niaz, 2008d, 2010a, 2011).

Liu and Tsai (2008) have reported that science major (pure science and science education) undergraduate students had less sophisticated beliefs with respect to the theory-ladenand cultural aspects of science, than non-sciencemajors. Interestingly, according to Bencze and Hodson (1999), in many cases the science curriculum itself acts as a barrier to change in teaching and understanding science. Future research could explore further the suggestion that it is precisely their beliefs about certainty and objectivity that lead students to select science as their major subject. At this stage, it is important to consider the advice of philosopher-physicistJames Cushing (1989), "One cannot simply amass a 'Baconian' heap of facts (or of theses) and then hope that truth or a theory will emerge" (pp. 19–20). Based on a historical reconstruction of science, Giere (2006, p. 5) advocates a "perspectival realism" (in contrast to objective realism), according to which no empirical claims can be known with absolute certainty to be true. In science education, Cobern and Loving (2008) have endorsed Giere's pluralistic view of progress in science, based on "perspectival realism."

Sadler et al. (2004) have argued cogently that science operates under the implicit assumption that scientific knowledge develops, builds upon itself, and changes over time. Furthermore, scientists would not devote their lives to the pursuit of knowledge if they had no chance of adding to, or changing, prevailing paradigms. Philosophers of science may wrangle over the mechanisms of scientific change (Kuhn, 1962; Lakatos, 1970; Laudan, 1977), but the manifestations of scientific tentativeness are relatively easy to incorporate in the classroom. This precisely constitutes a challenge to science educators, as both the textbooks and the curriculum ignore essential elements of the scientific endeavor, namely, alternative interpretations of data that lead to conflicts and controversies, and hence the tentative nature of science (Niaz, 2009a). At this stage a word of caution is necessary as we need to go beyond the rhetoric of tentativeness and fallibility by describing just when and how scientists and the scientific community were led into controversy and how it was resolved (Allchin, 2004b). This, of course, needs historical reconstruction of the different episodes that form part of the science curriculum.

In general, college students lack the understanding that theories are broader explanatory frameworks guiding hypothesis testing (Smith & Wenk, 2006) and this can perhaps be attributed, among other factors, to an authoritarian view of science (Marshall & Young, 2006), thus implying a more passive or rote learning. According to Grandy and Duschl (2007), the role of experiments is precisely situated in theory and model building, testing and revising (p. 143). It is plausible to suggest that the epistemic environment (theories and laws representing absolute truth) of science students and teachers, itself constitutes a hurdle in a better understanding of the NOS (Liu & Tsai, 2008). Rudolph (2008) considers such an environment as that of "science for scientists" (p. 65), in contrast to that of a "historical understanding" of science (p. 79). According to Grandy and Duschl (2007), in order to go beyond the "science for scientists" we need to emphasize, "how we know what we know and why we believe certain statements rather than competing alternatives" (p. 143). Collins (2007), for example, questions the presentation of the famous Michelson-Morleyexperiment in textbooks, as this is not how actual science is done. To illustrate how 'science in the making' is complex and controversial, Niaz (2009a) has presented various episodes from the history of science (including the Michelson-Morleyexperiment).

It is concluded that research reviewed in three major science education journals in the period 2004–2008 has focused more attention on identifying students' and teachers' epistemological beliefs (64%), whereas less work (36%) has been done on actual teaching strategies for facilitating understanding of the HPS. If we want teachers to understand a topic within an HPS perspective then we need to go beyond a few hours of NOS-relatedinstruction in a science-methodscourse (Abd-El-Khalick2005, p. 38). A major implication of this study is to stimulate research on facilitating students' and teachers' understanding of the HPS, based on classroom strategies for the different topics of the science curriculum. According to Clough (2006), NOS instruction would be more effective if teachers deliberately scaffold classroom experiences and students' developing NOS understanding back and forth along the continuum (p. 463). An example of such a teaching experience is provided by Niaz (2009b), in which participants were given the opportunity to reflect, discuss, and participate in a series of course activities based on various historical controversies that formed part of the science curriculum.

Finally, in my opinion, research on the NOS shows that it needs to be based on an HPS perspective in order to provide more thought-provokinginstances of 'science in the making.' In contrast, to the traditional science curriculum which invariably emphasizes "science for scientists," we need to ask, "Are we teaching science as practiced by scientists?" (cf. Niaz, 2010a). Historical reconstructions based on the work of modern philosophers of science show that experimental data do not unambiguously lead to the formulation of scientific theories (this may sound counter-intuitiveto most science teachers and students). A critical evaluation of the historical record reveals that scientists invariably interpret experimental data differently (due to the complexity of issues involved) and this leads to controversies and conflicts that require the participation of the scientific community. Science as practiced by scientists reveals that they are constantly engaged in a human enterprise based on presuppositions, contradictions, controversies, and speculations (Niaz, 2010b). It is concluded that teaching science as practiced by scientists would be more motivating for students and thus facilitate a better understanding of progress in science.

4

HOW TO INTRODUCE THE NATURE OF SCIENCE IN THE CLASSROOM

Introduction

The nature of science (NOS) has been the subject of considerable research in the science education research literature (Alters, 1997; Blanco & Niaz, 1997; Chinn & Brewer, 1993; Cobern, Gibson, & Underwood, 1999; Kang, Scharmann, & Noh, 2005; Lederman, Abd–El–Khalick,Bell, & Schwartz, 2002; McComas, 1996; Niaz, 2001b, 2008a; Osborne, Collins, Ratcliffe, Millar, & Duschl, 2003; Smith et al., 1997; Smith & Scharmann, 1999; Solomon et al., 1996). What exactly do we understand by the NOS? The essence of the NOS deals with how scientists do science (also see Chapter 1), namely, the:

a. relationship between experiment, data and theory;
b. role played by scientists' prior beliefs and presuppositions while they are designing new experiments;
c. interpretation of the same experimental data in different ways;
d. rivalries and conflicts among scientists as they conduct experiments and understand data;
e. continual critical appraisal of theories leads to the tentative nature of scientific knowledge;
f. objectivity in science as a social process of competitive cross–validation through critical peer evaluation (cf. Phillips & Burbules, 2000; Schwab, 1962).

An independent observer would respond that if this is how scientists work ('science in the making') then it should be a part of the science curriculum, text–books, and classroom activities. Nevertheless, the crux of the issue is that how sci–

entists do science cannot always be gleaned from the professional journal articles (cf. comments by Peter Medawar and discussion in Chapter 3). This, precisely, makes the task of introducing the NOS in the classroom difficult. To make matters worse both the science curriculum and the textbooks generally follow the inductivist–orientedaccounts presented in the professional journals.

Despite efforts by reform documents (AAAS, 1993), research based on textbook analyses shows that these and other aspects of the NOS are generally ignored in various topics of the science curriculum and the following are some of the examples:

a. The nature of science (Abd–El–Khalick,Waters, & Le, 2008, Niaz & Maza, 2011).
b. Scientific literacy themes (Chiappetta, Sethna, & Filman, 1991; Leite, 2002).
c. Normal science education (Van Berkel, De Vos, Verdonk, & Pilot, 2000).
d. Atomic structure (Justi & Gilbert, 1999; Niaz, 1998; Niaz & Coştu, 2009; Rodríguez & Niaz, 2002, 2004a).
e. Determination of the elementary electrical charge (Niaz, 2000b; Niaz & Rodríguez, 2005; Rodríguez & Niaz, 2004b).
f. The amount of substance and its unit the "mole" (Padilla & Furio–Mas, 2008).
g. Laws of definite and multiple proportions in chemistry (Niaz, 2001b).
h. The periodic table of chemical elements (Brito, Rodríguez, & Niaz, 2005).
i. Osmotic pressure (de Berg, 2006).
j. Heat and temperature (de Berg, 2008a).
k. Oxide chemistry (de Berg, 2008b).
l. Quantum mechanics (Niaz & Fernández, 2008; Shiland, 1997; Tampakis & Skordoulis, 2007).
m. Relativity theory (Arriassecq & Greca, 2007; Velentzas, Halkia, & Skordoulis, 2007).
n. The pendulum (Koliopoulos & Constantinou, 2005).
o. The photoelectric effect (Niaz, Klassen, McMillan, & Metz, 2010a).
p. Evolution (Skoog, 2005).

Research in science education has also shown considerable interest in designing teaching strategies for introducing the NOS in the classroom (Abd–El–Khalick & Akerson, 2004; Akerson, Morrison, & Roth McDuffie, 2006; Ford & Wargo, 2007; Khishfe & Lederman, 2006, 2007; Smith & Scharmann, 2008; Windschitl, 2004; Wong et al., 2008). Once again, there seems to be some consensus that teaching the NOS needs a reflective, explicit, and activity–based approach. According to Smith and Scharmann (2008): "NOS instruction is most likely to be effective when it requires active cognitive engagement with NOS issues and requiring that students reflect on current positions as well as changes in their understandings" (p. 229). It is important, however, to note that research in science education continues to debate the various strategies for introducing the NOS in the classroom (Chang, 2011; Hötteche, Henke, & Riess, 2011).

The objective of this chapter is to facilitate teachers' understanding of the NOS and explore the difficulties involved for its implementation in the classroom.

Rationale and Design of the Study

This study is based on 16 in-service chemistry teachers who had enrolled in the course, "Epistemology of Science Teaching," as part of a Master's degree program in education, at a major university in Latin America. Nine teachers worked in secondary schools and seven at the university level (male = 4, female = 12, age range: 25–40 years), and their teaching experience varied from about 5–15 years. In the previous year all teachers had enrolled in the following courses:

a. *Methodology of Investigation*, in which the basic philosophical ideas of Popper, Kuhn, and Lakatos were discussed, in order to provide an overview of the controversial nature of progress in science (growth of knowledge) and its implications for research methodology in education. Teachers were familiar that basic ideas like the scientific method, objectivity, and the inductive nature of science were considered to be controversial and questionable by philosophers of science.
b. *Investigation in the Teaching of Chemistry*, in which students' alternative conceptions and conceptual change strategies were discussed within a history and philosophy of science (HPS) perspective, with particular reference to historical controversies. For example, J.J. Thomson did not find E. Rutherford's interpretation of alpha particle experiments as logical, despite chemistry textbooks' claim to the contrary. Similarly, Bohr's postulation of the "quantum of action" was not accepted as a logical alternative by many renowned physicists (cf. Niaz, 1998; this was one of the readings in this course). The Millikan–Ehrenhaft controversy with respect to the oil drop experiment was another issue discussed in this course. This provided evidence for alternative interpretations of similar experimental data (Duhem–Quine thesis) and was discussed in one of the readings (Niaz, 2000b). This course facilitated participants' understanding with respect to the tentative nature of science and that progress in science inevitably involves rivalries among different interpretations/research programs.

Course content (Reading list)

The course was based on 13 required readings and was subdivided into the following units:

- *Unit 1: Nature of Science*: (1) Campanario (1999); (2). Mellado (2003); (3a) Niaz (2001a); (3b) Scharmann & Smith (2001); (4) Fernández, Gil, Carras–cosa, Cachapuz, & Praia (2002).
- *Unit 2: Critical Evaluation of Nature of Science*: (5) Kang, Scharmann, & Noh (2005); (6) Petrucci & Dibar (2001); (7) Khishfe & Lederman (2006); (8) Akerson, Morrison, & Roth McDuffie (2006); (9) Windschitl (2004).
- *Unit 3: Critical Evaluation of Constructivism*: (10) Marín, Solano, & Jiménez (1999); (11) Martínez (1999); (12) Gil–Pérezet al. (2002); (13) Niaz et al. (2003).

Course Organization and Activities

On the first day of class (2 hours) all participants were provided copies of all the readings and salient features of the course were discussed. It was emphasized that the course called for active participation. As all teachers worked in nearby schools and universities, three types of course activities were programmed:

a. Class discussions were planned on Saturdays of the 6th and 8th weeks of the course (three hours in the morning and three in the afternoon). Read–ings 1–7 were discussed in the first meeting and readings 8–13 in the second meeting. Teachers were supposed to have studied each of the readings before the meetings. Each meeting started off with various questions and comments by the participants. The instructor intervened to facilitate understanding of the issues involved (total time devoted to class discussions = 12 hours).
b. Class presentations by the participants were programmed during the 11th and the final week of the course (Monday to Saturday, total time = 44 hours). On the first day of class, all participants selected (by a draw, some in groups) one of the 13 readings for a presentation. Each participant was assigned 90 min–utes (30 minutes for the presentation and 60 minutes for interventions and discussions). Each of the presentations was moderated by one of the partici–pants. The instructor intervened when a deadlock was reached on an issue. It was expected that the participants would present the important aspects of the readings, with the objective of generating critical discussions. All participants prepared PowerPoint presentations.
c. During class presentations in the final (11th) week, participants were encour–aged to ask their questions in writing, which were then read out loud by the moderator. The presenters were given the opportunity to respond and then a general discussion followed. At the end of each session all written questions were submitted to the instructor, which provided important feedback with respect to issues, conflicts, and interests of the participants. Each participant signed her/his question. The same procedure was followed in all 13 presenta–tions, generating a considerable amount of data (a pool of 250 questions and comments). Besides this data, the instructor also took class–notesthroughout

the course. This data facilitated the corroboration (triangulation of data sources) of participants' responses to the seven exam questions in this study.

Course Evaluation

All participants presented an Initial exam in the first session and a Final exam during the last session of the final week (11th). The rationale behind these evaluations in this study was not merely evaluation in the traditional sense. Both exams were "open book" (about three hours each) and participants were allowed to consult any material that they felt could be helpful. The Initial exam reflected participants' experience during the first 10 weeks, including 12 hours of class discussions. The Final exam reflected an additional experience of 36 hours of class presentations and discussions during the final week. Furthermore, both exams were based on the premise that oral presentations and discussions can be complemented by written responses and thus facilitate greater understanding. This study is based on participants' written responses to one item of the Initial exam and three items of the Final exam. These four items were formulated in order to provide possible answers to the objectives of this study. The four items included in the Initial and Final exams are presented below:

- *Initial exam*
  - *Item 1a*: What do you understand by the nature of science?
  - *Item 1b*: Select four aspects of the nature of science that in your opinion are the most important for teaching chemistry.
- *Final exam*
  - *Item 2a*: Why is it difficult to achieve consensus with respect to teaching the nature of science?
  - *Item 2b*: Select one aspect of the nature of science that in your opinion is the most important and how would you incorporate it in your classroom?
  - *Item 3*: Based on your understanding of the nature of science, would you agree or disagree with the following affirmations (please explain):
    a. Based on experimental data, scientific theories emerge in a logical manner.
    b. The function of a hypothesis is to foresee experimental results and not to facilitate an explanatory framework.
    c. Accumulation of experimental data is the most important aspect in the development of a theory/law.
  - *Item 4*: Many investigators emphasize the role played by experimental data in scientific development. In contrast, some other investigators emphasize the role played by presuppositions (theoretical framework of the scientist). Consider the following three models for understanding

scientific development presented in Reading 9 (Windschitl, 2004): Model 1, Figure 3, p. 505; Model 2, Figure 4, p. 506; and Model 3, Figure 5, p. 507. In your opinion, which of these models best explains scientific development (Please explain).

Multiple Data Sources

Based on the different course activities this study generated the following data sources:

a. Question–answer sessions after each of the 13 formal presentations, in which participants wrote their questions/comments (interventions), which were then discussed in class and later submitted to the instructor (a pool of 250 questions and comments were generated).
b. Initial and Final exams during the final week, separated by 36 hours of class presentations and discussions (total time = 8 hours).
c. Questionnaire based on two items (Q1 and Q2) that was responded to on the third day of the final week, as a take home activity.
d. The instructor's class–notes,based on the following activities throughout the course: class discussions during the 6th and 8th weeks of the course (total time = 12 hours), 13 formal presentations during the final week, question–answer sessions after each presentation. These notes consisted of quotes from students' (including their names) interventions during presentations and discussions, accompanied by general comments.
e. Critical essay. As part of their evaluation all participants were required to submit a critical essay based on any one or various readings. The objective of this essay was to present a critique based on epistemological, philosophical, and methodological aspects. This essay was submitted 10 days after the Final exam, which provided the participants ample time to reflect and elaborate their ideas.

At this stage it is important to emphasize the role played by multiple data sources in educational research. (Note: Teachers in this study participated in 56 hours of various class activities.) Given the nature of the paradigm wars (Gage, 1989; Howe, 1988), educational literature has suggested the need to move beyond the quantitative versus qualitative research designs and called for mixed methods research (Guba & Lincoln, 1989, 2005; Johnson & Onwuegbuzie, 2004; Niaz, 2011; Shulman, 1986).

Issues Related to the Nature of Science Explicitly Discussed in Class

In order to facilitate participant teachers' understanding, various aspects of the NOS in the following readings were discussed:

a. Campanario (1999), Reading 1, critiques the science curriculum, textbooks, and the popular science culture, which present an immaculate vision of science in which scientists work tirelessly and disinterestedly to help humanity by presenting the "truth."
b. Mellado (2003), Reading 2, establishes an analogy between science teachers' conceptual change and various models of progress in science, such as positivism, Popper, Kuhn, Lakatos, Toulmin, and Laudan.
c. Niaz (2001a), Reading 3a, critiques Smith and Scharmann (1999) and suggests that the following aspects of the NOS can also be included: Competition among rival theories, the same experimental data can be interpreted in more than one way (theory ladenness of observations), and the inconsistent foundation of some scientific theories. Smith and Scharmann (1999) had previously drawn attention to the importance of the NOS for science teachers and suggested the following as essential for understanding progress in science: Science is empirical, scientific claims are testable/falsifiable, science is tentative and self-correcting.
d. Scharmann and Smith (2001), Reading 3b, is a response to Niaz (2001a) and facilitates further understanding of the NOS. This provided participants an opportunity to understand how researchers in science education can disagree with respect to how the NOS can be introduced in the classroom.
e. Fernández, Gil, Carrascosa, Cachapuz, & Praia (2002), Reading 4, reviewed science education literature over a period of 10 years (1990–1999), in the following journals: *Science Education, International Journal of Science Education, Journal of Research in Science Teaching, Studies in Science Education, Science & Education*, and *Enseñanza de las Ciencias*. This review revealed the following deformed conceptions with respect to the NOS: (1) Empirico-inductivistand atheoretical; (2) Rigid, algorithmic, and infallible; (3) Aproblematic, ahistoric, and dogmatic; (4) Exclusively analytic based on simplification; (5) Accumulative and lineal growth; (6) Individualist and elitist; and (7) Decontextualized and socially neutral. An important aspect of this reading was that it included the references to all the research articles on which the study was based, thus providing a considerable amount of evidence from the research literature.
f. Kang, Scharmann, & Noh (2005), Reading 5, studied Korean students' views of the NOS with respect to the following aspects: the purpose of science, the definition of scientific theory, the nature of models, the tentativeness of scientific theory, and the origin of scientific theory. Results indicated that the majority of students possessed an absolutist/empiricist perspective about the NOS.
g. Petrucci and Dibar (2001), Reading 6, studied understanding of the NOS by students in Argentina. This study reported that students have considerable difficulty in understanding the following NOS-relatedaspects: What are the objectives of science? Do scientific theories change? How do we know if a piece of knowledge is scientific or not?

h. Khishfe and Lederman (2006), Reading 7, investigated the influence of two different explicit instructional approaches in facilitating more informed understanding of the NOS among 9th grade students in Chicago. In the "integrated" group, NOS instruction was related to the science content about global warming. For the "non–integratedgroup," the NOS was taught through a set of activities that specifically addressed NOS issues and were dispersed across the content about global warming. Results showed improvements in students' views of the NOS regardless of whether the NOS was integrated or not.
i. Akerson, Morrison, and Roth McDuffie (2006), Reading 8, investigated the retention of views of the NOS by graduate pre–serviceteachers after having received explicit reflective instruction. After five months teachers were again interviewed and it was found that several of them reverted to their earlier views. Authors concluded that one course is not enough for teachers to accept meaningful changes in their NOS views.
j. Windschitl (2004), Reading 9, reported a study in which pre–servicesecondary science teachers developed their own empirical investigations, from formulating questions to defending results in front of peers. Findings indicated that participants shared a tacit framework of what it means to "do science" which shaped their investigations. However, the most consistent assumptions were misrepresentations of fundamental aspects of science, such as: (1) A hypothesis functions as a guess about an outcome, but is not necessarily part of a larger explanatory system; (2) Background knowledge may be used to provide ideas about what to study, but this knowledge is not in the form of a theory or model; and (3) Theory is an optional tool one might use at the end of a study to help explain results. The author concluded: "These ideas appear consistent with a 'folk theory' of doing science that is promoted subtly, but pervasively, in textbooks, through the media, and by members of the science education community themselves" (p. 481).

This synopsis of the course content shows quite clearly how participant teachers were constantly provided feedback with respect to the NOS and difficulties involved for its implementation in the classroom. Furthermore, teachers participated and became aware of the continuing debate on these topics.

Results and Discussion

In this section participants' responses to the four evaluation items (presented in the previous section) are presented. In each of the items all responses were analyzed and, based on this, various categories were generated. Each of the responses was classified into one of the categories. Some of the responses could easily have been classified in more than one category.

Understanding the Nature of Science (Participants' Responses to Item 1a)

Item 1a asked participants to consider and reflect on the following: What do you understand by the nature of science? The following responses illustrate participants' thinking:

> The nature of science comprises not only the discoveries, but rather how and through what contingencies and what were the underlying reasons for undertaking research. All these aspects form a complex structure that is necessary if we want to understand and develop science. *(Participant #1)*

> It is a reflective and metacognitive process with respect to what is science. It means that we have to understand not only the content of science but also its origin, which is impregnated with prior conceptions and a particular social context. *(Participant #6)*

> The nature of science represents the "real" characteristics of science, namely how and why we do science—very briefly it represents "human nature." In my understanding these characteristics are not exclusive of the scientific enterprise, but rather common to various human endeavors. *(Participant #11)*

> The nature of science can be conceived as a progressive transition of heuristic principles (Schwab, 1962, 1974). In other words, progress in science does not depend exclusively on experimental data but rather on the ability of the scientist to analyze and interpret based on the underlying heuristic principles. *(Participant #12)*

> The nature of science involves epistemological, philosophical and heuristic aspects that help to understand scientific development. It includes characteristics of the researcher, her/his beliefs, theoretical ideas that could have influenced the development of scientific activities, possible errors and other unforeseen situations that could influence the results and their interpretations. *(Participant #15)*

These responses show that participants considered the NOS to be a complex process based on various contingencies, which are in turn determined by epistemological, philosophical, and heuristic aspects of the problem situation. Furthermore, "doing science" being a human enterprise it involves difficulties that are similar to everyday life problems.

Important Aspects of the Nature of Science (Participants' Responses to Item 1b)

Item 1b asked participants to select four aspects of the NOS that are most important for teaching chemistry. Based on the participants' responses, the categories shown in Table 4.1 were generated.

1. The Tentative Nature of Scientific Theories (n = 13)

The following responses illustrate participants' thinking:

> History of science shows that in general a theory is substituted by another and this in turn is replaced by another—however theories are never definitive. For example, in chemistry the laws of definite and multiple proportions were substituted by atomic theory, which in turn has gone through many changes and will continue to do so. Furthermore, each successive theory provides greater explanatory power.
>
> *(Participant #1)*

> In the presence of new evidence, scientific knowledge can change. Scientific theories always change, which could be progressive or regressive similar to research programs (Lakatos, 1970). Similarly, Kuhn (1970) contemplates changes in scientific theories, which can be radical or revolutionary.
>
> *(Participant #6)*

> Scientific theories are accepted so long as these can explain most of the phenomena of interest to the scientific community. This shows the progressive and dynamic nature of science and that "scientific truths" are transitory. However, when new evidence appears that contradicts a theory, the old theory progressively gets abandoned. This precisely is the difference

TABLE 4.1 Participants' Responses to Item 1b (Select four aspects of the nature of science that in your opinion are the most important for teaching chemistry)

*Response category**	*Frequency (n)*
1. Tentative nature of scientific theories	13
2. Observations are theory laden	9
3. Role of conflicts, controversies and rivalries in scientific progress	8
4. Scientists use their imagination and creativity to develop theories	7
5. Scientific knowledge relies on experimental evidence, rational arguments and skepticism	6
6. Objectivity in science is a consequence of a social process	5
7. There is no universal scientific method	4
8. Science is self-correcting	4

* These categories were generated based on participants' responses to Item 1b.

> between science and religion. In religion we have dogmas: eternal truths. In contrast, science is a construction of "reality." *(Participant #8)*

> Many of the concepts that we teach in the classroom are not "absolute truths" but, rather, "ideal models" (heuristic principles) with respect to a particular phenomenon, which are adequate for a particular time and permit scientific advance. However, these models are susceptible to modification and even perhaps elimination as the new investigations advance. *(Participant #11)*

Participant #1 has referred to the relationship between the laws of definite and multiple proportions and atomic theory, which has been the subject of considerable research in the HPS literature (Christie, 1994; Rocke, 1978). As suggested by this participant the transition from the law of definite proportions to its explanation by the atomic theory is correct. However, the relationship between the law of multiple proportions and atomic theory is more complex as it was Dalton who, based on his atomic theory, predicted the law (cf. Niaz, 2001b for details). Even, Linus Pauling (1964) although no philosopher of science, recognized this. Overall, participants in this study grasped the basic idea of scientific theories being tentative and never definitive. Furthermore, scientific theories do not represent "absolute truths" but rather "ideal models" based on heuristic principles. As a consequence if heuristic principles change (among other reasons due to new evidence), so would the models. It is important to note that heuristic principles may also change as another scientist may come up with a better theoretical framework to explain the same experimental data. One of the participants referred to the important distinction between religion as dogma and science as construction.

2. *Observations are Theory Laden (n = 9)*

The following responses illustrate participants' thinking:

> Every scientist has a prior idea as to how the experiment would run and the expected results. In the majority of the cases the scientist expects to verify what she/he had in mind. Consequently, scientists defend data that support their theories and thus provide evidence for a particular point of view. The theory–ladennature of observations, help us to understand better, for example the experimental work of J.J. Thomson and E. Rutherford. To a certain extent, it is good that scientists defend their criteria. However, at the same time they must have an open mind, for possible new alternatives. This also shows that scientists are not "objective" in an absolute sense. *(Participant #1)*

> As observations are theory laden, besides having experimental data what is important are the interpretations based on heuristic principles, which are in turn based on arguments and critical appraisals. *(Participant #8)*

> In order to "construct" knowledge it is not sufficient to simply do the experiments. All experiments have an underlying heuristic principle or pre-suppositions (some quite audacious), which help in the interpretation of the data. In the context of the "discovery" of the electron the experimental data of W. Kaufmann and E. Wiechert, as compared to that of J.J. Thomson is a good example of how observations are theory laden. *(Participant #11)*

These responses clearly emphasize the role of the theory-laden nature of observations and that it is precisely the relationship between heuristic principles and experimental data that help us to understand science. Participant #1 has referred to the work of J.J. Thomson and E. Rutherford, in the context of atomic models, which was the subject of a reading in a previous course (Investigation in the teaching of chemistry, cf. section "Rationale and Design of the Study"). To explain the findings of Rutherford's alpha particle experiments, Thomson was led into a bitter dispute with his student and colleague. To maintain his model of the atom and to explain large-angle deflections of alpha particles, Thomson put forward the hypothesis of compound scattering, whereas Rutherford had explained the same experimental finding through his hypothesis of single scattering (for details, see Niaz, 2009a, Chapter 6). Participant #11 has compared the experimental work of Kaufmann and Wiechert, with that of Thomson, which was the subject of a previous course (Investigation in the teaching of chemistry, cf. section "Rationale and Design of the Study"). Determination of the mass-to-charge ratio (m/e) of the cathode rays can perhaps be considered the most important experimental contribution (based on a heuristic principle) of Thomson's (1897) article. Yet, he was neither the first to do so nor the only experimental physicist. Two German physicists, Kaufmann and Wiechert, also separately determined (m/e) ratios of cathode rays in the same year as Thomson and their measurements agreed with each other. Falconer (1987) explained cogently how Thomson's contribution differed from that of Kaufmann and Wiechert:

> Kaufmann, an ether theorist, was unable to make anything of his results. Wiechert, while realizing that cathode ray particles were extremely small and universal, lacked Thomson's tendency to speculation. He could not make the bold, unsubstantiated leap, to the idea that particles were constituents of atoms. Thus, while his work might have resolved the cathode ray controversy, he did not "discover the electron." *(p. 251)*

Apparently, Thomson's ability to speculate, elaborate alternative hypotheses and models, and, perhaps most importantly, formulate a theoretical framework for his experimental findings, led him to foresee and conceptualize what his contemporaries ignored.

3. The Role of Conflicts, Controversies, and Rivalries in Scientific Progress (n = 8)

The following responses illustrate participants' thinking:

> Studying science contextually with reference to actual historical episodes shows that scientists put forward varying interpretations that are impregnated with their respective theoretical frameworks, leading to conflicts and controversies. *(Participant #6)*

> If scientists can have different heuristic principles to interpret the same experimental data, then critical appraisals, arguments and varying interpretations inevitably lead to controversies and rival theories. These contradictions are then subjected to peer evaluation, which helps to establish the "objective" nature of scientific theories. *(Participant #8)*

An important aspect of these responses is the need to study and teach science in the context in which it developed. If a topic forms part of the science curriculum students would be more receptive to follow its historical reconstruction and, consequently, understand the NOS. This precisely shows that progress in science inevitably involves difficulties leading to controversies among rival theories.

4. Scientists Use their Imagination and Creativity to Develop Theories (n = 7)

The following responses illustrate participants' thinking:

> In the early phases of model building scientists do not have all the necessary pieces of evidence and hence they have to resort to their imagination and creativity. Different models of the atom provide a good example to illustrate this aspect. *(Participant #1)*

> Based on their heuristic principles, scientists construct models that provide possible explanations of phenomena. These models can vary from scientist to scientist and hence require imagination and creativity to play an important role. *(Participant #3)*

These responses recognize the role played by heuristic principles in the construction of models, especially in the early phases and hence scientists often resort to their imagination and creativity.

5. Scientific Knowledge Relies on Experimental Evidence, Rational Arguments and Skepticism (n = 6)

The following response illustrates participants' thinking:

> Our present understanding of science questions the traditional view of science in which experimental evidence was the sole arbiter in the elaboration of a theory. My view is illustrated by the following examples from the history of science: Big bang theory, sub–atomicparticles, and determination of the position of an electron. *(Participant #8)*

Participants' thinking in this category coincides with the characterization of the NOS in various reform documents (cf. McComas, Almazroa, & Clough, 1998). A historian of science has recently expressed this aspect in cogent terms:

> Tracking the history of the Universe from the instant after the Big Bang is a stellar achievement—but it leaves unanswered the fundamental question of how the Big Bang started in the first place. As so often in science, how you interpret the evidence depends on how you want to see it. *(Fara, 2009, p. 394)*

6. Objectivity in Science is a Consequence of a Social Process (n = 5)

The following response illustrates participants' thinking.

> It is well known that progress in science has frequently involved controversies among rival theories. These contradictions among scientists are then subjected to peer review and it is precisely this validation that facilitates objectivity in science. *(Participant #8)*

This response constitutes an important reflection on scientific research methodology, namely, how are contradictions or conflicts resolved. Campbell (1988a), a methodologist has explained this in the following terms: "the objectivity of physical science does *not* come from the fact that single experiments are done by reputable scientists according to scientific standards. It comes instead from a social process which can be called competitive cross–validation"(pp. 302–303, original italics).

7. There is no Universal Scientific Method (n = 4)

The following responses illustrate participants' thinking:

> The simplicity with which the scientific method is presented in textbooks and the classroom, obscures the complexity of the methodological strategies used by scientists. *(Participant #6)*

> The research methodologies used by Millikan and Ehrenhaft to determine the elementary electrical charge help us to understand the lack of a universal scientific method. Ehrenhaft followed the scientific method rigorously by letting the experimental data dictate his theory. In contrast, Millikan was guided by his presuppositions and selected data carefully in order to corroborate his theoretical framework. *(Participant #8)*

Despite the popularity and extensive use of the scientific method in textbooks and the classroom, it has been questioned by philosophers of science and also science education researchers. The Millikan–Ehrenhaft controversy and their respective methodologies was the subject of a reading in a previous course (Investigation in the teaching of chemistry, cf. section "Rationale and Design of the Study"). According to Niaz (2009a):

> It is plausible to suggest that Ehrenhaft's methodology approximated to the traditional scientific method which did not allow him to discard "specious drops". Millikan, on the other hand, in his publications espoused the scientific method, but in private (cf. hand–writtennotebooks) was fully aware of the dilemma faced and was forced to select data in order to uphold his presuppositions. *(p. 117)*

8. Science is Self-correcting (n = 4)

The following response illustrates participants' thinking.

> Students are generally not aware that all scientific theories can potentially be refuted. For this reason all research must be confronted with critical appraisals based on peer review. It is the scientific community that ensures the self–correctingaspect of science through publication in peer reviewed journals. *(Participant #11)*

This response shows how science monitors its own progress through the peer review evaluation process. In contrast, both textbooks and teachers generally ignore that even famous scientists have had their work subjected to critical appraisals.

Consensus in Teaching the Nature of Science (Participants' Responses to Item 2a)

Item 2a, asked participants: Why is it difficult to achieve consensus with respect to teaching the nature of science? The following responses illustrate participants' thinking:

> Achieving consensus is difficult due to many parallel lines of research. Some of these emphasize entirely empirical perspectives, whereas others are more interpretative based on heuristic principles (Schwab, 1962). *(Participant #9)*

> In my opinion, there are various reasons for this: a) For most teachers teaching science means solving algorithmic problems. Not long ago, I also agreed with this positivist perspective in which the historical origin of a theory or idea is ignored; and (b) In-serviceand pre-serviceteachers are not exposed to critical reflections with respect to science content. *(Participant #10)*

> Teaching nature of science (NOS) is a complex task and the research literature also shows that it is difficult to achieve consensus. For example, Scharmann and Smith (2001, Reading 3b) have suggested that some NOS aspects are too complex to be taught at the secondary level. On the contrary, Niaz (2001a, Reading 3a) would recommend a wider spectrum of NOS aspects to be included. Furthermore, the influence of positivism/empiricism is deeply rooted among teachers, which makes any change not only difficult but also polemical. *(Participant #11)*

> Teaching science is by itself quite difficult. Besides if we include nature of science (NOS), which deals with polemics and controversies during the development of a theory/law (e.g., Millikan's oil drop experiment), the classroom may turn into chaos and some students may lose motivation. However, if these controversies are well administered a consensus can be achieved. Furthermore, inclusion of NOS ultimately would depend on teacher training. *(Participant #12)*

> Any consensus would ultimately depend on the theoretical referent and the epistemological beliefs of the teachers and textbook authors. For most teachers science is a finished product with no flaws or deficiencies and hence any change is difficult. Before achieving consensus, the new idea will have to first face a conflict with the dominant positivist paradigm and then hopefully it may prevail. *(Participant #13)*

It seems that for most participants introduction of the NOS based on a postpositivist (Phillips & Burbules, 2000) vision of science is difficult as it conflicts with the dominant positivist paradigm. Furthermore, the new vision of the NOS also contradicts some of the cherished objectives of science courses, such as: solving algorithmic problems, presenting science as a finished product with no deficiencies, and lack of a historical perspective based on controversies and other aspects of the NOS. Participants also drew attention to the difficulties involved in introducing the NOS, such as teacher preparation and student motivation.

Some participants referred to the differences among science education researchers themselves (e.g., Readings 3a & 3b).

Incorporating the Nature of Science in the Classroom (Participants' Responses to Item 2b)

Item 2b asked participants to: Select one aspect of the nature of science that in your opinion is the most important and say how you would incorporate it in your classroom.

Based on participants' responses the categories shown in Table 4.2 were generated.

1. The Tentative Nature of Scientific Theories (n = 7)

The following responses illustrate participants' thinking:

> This NOS aspect helps us to understand that the "definitive" in science has not been found and hence we have to keep investigating so that we and our students can also make some contribution toward scientific progress. In order to introduce this aspect in the classroom I would select a controversial topic from the history of science, such as the atomic models of Thomson and Rutherford or perhaps a more recent topic of string theories. This will make students understand that in science we shall always be looking for the "truth" or perhaps "half truth." *(Participant #1)*

> This will help to show to the students that science was not created by geniuses to find "absolute truths," but is subject to change and improvement. Based on the historical details most topics of the curriculum can be adapted for implementing NOS. *(Participant #2)*

> Scientific knowledge is not a "finished" product but rather subject to constant change. The changing nature of atomic models is a good example to introduce in the classroom. *(Participant #6)*

TABLE 4.2 Participants' Responses to Item 2b (Select one aspect of nature of science that in your opinion is the most important and how would you incorporate it in your classroom?)

*Response category**	*Frequency (n)*
1. Tentative nature of scientific theories	7
2. Role of heuristic principles: Observations are theory laden	5
3. Competition among rival theories: Same experimental data can be interpreted differently	4

* These categories were generated based on participants' responses to Item 2b.

> This aspect is like a transverse axis that can be adapted in many topics of the curriculum. Furthermore, it is important to show to the students that scientists are not elected by the gods and hence are fallible. Consequently, students can also become eminent scientists. *(Participant #8)*

> The periodic table of chemical elements is a good example to illustrate this aspect in the classroom as it has been changing and improving over many decades. Even the new chemical elements have been accommodated in the table. *(Participant #16)*

Most participants recognized that inclusion of the tentative nature of science in the classroom can help students to understand that "truth" in science is always elusive and this facilitates a constant process of changing and improving scientific theories. Among other topics, the changing nature of atomic models was considered to be a good example for students. Furthermore, as scientists are neither gods nor geniuses, the tentative nature of science means that our students can also contribute towards scientific development.

2. The Role of Heuristic Principles: Observations are Theory Laden (n = 5)

The following responses illustrate participants' thinking:

> All investigations, discoveries, theories and laws are guided by the heuristic principle of the researcher. Consequently, science is not an aggregate of "truths" but rather a progressive transition of heuristic principles. In the classroom I will try to focus on how scientists can turn out to be "wrong" and still promote scientific advance. Once the student comes to understand the "real and human" face of science it would be easier to implement other aspects of NOS. *(Participant #11)*

> Despite the importance of experimental data it is the heuristic principles that help scientists to interpret, elaborate theories and thus facilitate progress in science. For example, while teaching atomic structure it is important to present not only the atomic model but also the details and ideas that helped to elucidate data. *(Participant #12)*

These responses emphasized the role played by heuristic principles as guiding assumptions (Laudan, Laudan, & Donovan, 1988), underlying experimental findings. Furthermore, if a scientific theory is changed or even replaced by another, it does not mean that the earlier scientific endeavor was "wrong" or fruitless. This precisely shows that progress in science is a human enterprise (Niaz, 2009a).

3. Competition among Rival Theories: The Same Experimental Data can be Interpreted Differently (n = 4)

The following responses illustrate participants' thinking:

> Different research programs facilitate the formulation of rival scientific theories that compete among themselves (Lakatos, 1970). In the classroom we can design activities that can lead to the following: there is no absolute "truth," scientific method is questionable, and scientists are frequently involved in polemics and controversies. This background can produce cognitive conflicts among students and thus facilitate conceptual change. In the long run students would understand that based on the dedication of the researchers, scientific theories can evolve for decades and are not the work of magicians who discover the "truth." *(Participant #9)*

> Understanding that a variety of interpretations can be put forward to explain the same experimental data deepens our knowledge with respect to a problem under investigation. For example, in the case of the periodic table I would first present to the students some of the experimental data that were available in the 1860s. As a next step, I would ask students to analyze the data and discuss among themselves. I would then present the periodic tables of Mendeleev, Meyer and others in order to show their differences. Finally, I would explain how the periodic law was changed and the modern periodic table based on it. *(Participant #13)*

The role of cognitive conflicts as a teaching strategy in the classroom has been recognized in the science education literature and can be adapted for teaching various aspects of the NOS (Niaz, 1995a; Zohar & Aharon–Kravetsky, 2005). Questioning the scientific method and understanding that scientists try to explain phenomena and are not necessarily looking for "truth," is important if we want students to understand that interpretation of data in the history of science has been difficult and has often led to disputes and controversies. According to Leon Cooper (Nobel laureate in physics, 1972), varying interpretations of observations do not undermine the objective nature of science, but rather facilitate greater understanding (cf. Niaz, Klassen, McMillan, & Metz, 2010b).

Scientific Theories Emerge in a Logical Manner (Participants' Responses to Item 3a)

Item 3a required participants to agree/disagree with the following affirmation: "Based on experimental data, scientific theories emerge in a logical manner." The rationale behind this item was to provide participants a categorical statement that reflects the logical positivist perspective, and evaluate the degree to which they agree/disagree. All the participants except one disagreed with this affirmation and the following are some of the responses that illustrate participants' thinking:

> I disagree as sometimes ideas do not emerge from data but are already present in the minds of the scientists, even before doing the experiments. Sometimes, a theory may emerge from data but not necessarily in a logical manner as observations are always impregnated with subjectivity. *(Participant #1)*

> I disagree as it is not the experimental data but the scientific community that in the final analysis validates scientific theories. Many scientists create knowledge based on their imagination and creativity and then extrapolate to provide explanations that are not directly testable. A good example was Einstein who created theories with very little experimentation. *(Participant #2)*

> I disagree as this affirmation is taking into consideration only experimental data and neglecting the interpretations based on heuristic principles. One example is provided by the determination of the elementary electrical charge, in which Millikan did not "design" the experiment but "discovered" it (Holton, 1978). In other words, Millikan had a prior belief that the elementary electrical charge (heuristic principle) existed, and this led him to the need for experimental verification. *(Participant #9)*

> If this affirmation were correct then different scientists would interpret experimental data in the same manner and reach the same conclusions. History of science, however, provides multiple examples of how different researchers interpret the same data differently, and construct theories that are diametrically contrasting and compete to explain the same phenomena. A good example is the controversy between J.J. Thomson and E. Rutherford. *(Participant #11)*

Participants who disagreed explicitly referred to the theory-laden nature of observations and the difficulties involved in interpreting data which requires imagination and creativity. Furthermore, these participants provided various examples from the history of science, such as: Einstein, Millikan, Thomson, and Rutherford. The distinction between the "design" and "discovery" of the oil drop experiment is quite illustrative of the dilemma involved in the empiricist perspective (cf. Holton, 1978, pp. 184–185). The experiment itself could not have been performed (design) without alluding to the electron theory, that is, Millikan's guiding assumptions. Millikan's contribution consisted precisely in having "discovered" the experiment, which is recognized in succinct terms by Nagel (1961):

> It is unlikely that Millikan (or anyone else) would have devised [discovered] the oil-dropexperiment if some atomistic theory of electricity had not first suggested a question that seemed important in the light of the theory and that the experiment was intended to settle. *(p. 90)*

Similarly, the controversy with respect to the interpretation of alpha particle experiments between J.J. Thomson and E. Rutherford illustrates an important facet of how Thomson refused to accept Rutherford's interpretation as a logical consequence of the alpha particle experiments. Thomson propounded the hypothesis of compound scattering whereas Rutherford supported the hypothesis of single scattering as a possible explanation of the same experimental data (for details, see Niaz, 2009a, Chapter 6).

The only participant who agreed with this affirmation expressed his thinking in the following terms:

> I agree that the way theories emerge is primarily a logical process and I consider my position to be: positivist-constructivist.It is positivist as I consider the data to be very important in the formulation of theories. At the same time I am also a constructivist as data need interpretations that lead to contradictions and rivalries. *(Participant #4)*

This response reflects the underlying tension between the positivist and postpositivist (could be constructivist) conceptualizations of scientific progress. Indeed, during class discussions various participants referred to their previous understanding of science as being entirely positivist.

Hypothesis as an Explanatory Framework (Participants' Responses to Item 3b)

Item 3b required participants to agree/disagree with the following affirmation: "The function of a hypothesis is to foresee experimental results and not to facilitate an explanatory framework." This was adapted from Windschitl (2004, p. 491) who considered this facet to be a misrepresentation of one of the most fundamental aspects of scientific inquiry among pre-service secondary science teachers. Four participants agreed with this affirmation and 12 disagreed. The following are examples that illustrate responses of participants who disagreed:

> When a scientist formulates a hypothesis, she/he already has an idea with respect to the expected results and its explanation, and if these do not coincide then generally a new hypothesis is formulated as a function of results obtained. Consequently, a hypothesis does facilitate an explanatory framework. *(Participant #1)*

> Based on his/her previous experience and explanatory framework, a scientist formulates a hypothesis. Thus the two aspects (results and explanation) are included in a hypothesis. *(Participant #5)*

> A research program generates a hypothesis which is based on a heuristic principle that manifests a subject of interest for the scientist. Consequently, a hypothesis not only helps to conceptualize a problem but also explain experimental data. *(Participant #11)*

> A hypothesis precisely constitutes preconceived ideas or presuppositions of the scientist with respect to the expected results. Consequently, it has an explanatory function. *(Participant #13)*

These responses show that the participants clearly understood the dual role of a hypothesis, namely to foresee the results and provide a tentative explanation. Furthermore, the elaboration of a hypothesis is itself a complex process based on the scientists' previous experience, presuppositions, and heuristic principles. In other words a hypothesis is not a wild guess but, rather, a well thought out possible explanation.

The following is an example of a response from a participant who agreed with this affirmation:

> The primary function of a hypothesis is to foresee the results of an experiment. The explanatory framework is formulated later based on the results obtained. *(Participant #4)*

This response lacks the understanding that in a scientific inquiry a hypothesis is, at the same time, a tentative explanation. Of course, based on the results the tentative explanation could be further elaborated into a framework.

The Role of Experimental Data in the Development of a Theory/Law (Participants' Responses to Item 3c)

Item 3c required participants to agree/disagree with the following affirmation: "Accumulation of experimental data is the most important aspect in the development of a theory/law." Interestingly all participants disagreed with this affirmation and the following are examples that illustrate participants' thinking:

> I am in total disagreement as a simple accumulation of data conduces to nothing. The most important aspect in the development of a theory is the "heuristic principle" (Schwab, 1962), which provides the interpretation. Furthermore, the formulation of robust theories does not necessarily require many data. For example, Einstein postulated important theories with very little data or none at all. *(Participant #1)*

> Theories do not originate in a linear fashion based on the accumulation of data that is like a "Baconian inductive ascent" (Lakatos, 1970). On the contrary, science is not entirely empirical as it requires imagination and creativity and often leads to controversies. *(Participant #2)*

> I disagree as Millikan had to discard a major part of his data in order to postulate his theory, namely determination of the elementary electrical charge. In the development of a theory the most relevant aspects are: interpretation of the data based on researchers' perceptions and arguments and above all a consensus in the scientific community. *(Participant #5)*

> I disagree as accumulation of data can stay within the four walls of the lab. What really makes the difference is the discussion in the science community based on peer review. For example, in the case of the law of multiple proportions it was first proposed by Dalton based on his atomic theory and later verified by experiment. *(Participant #9)*

A majority of the participants emphasized the role played by the heuristic principles in the interpretation of experimental data. Participants provided examples from the history of science to corroborate this perspective (e.g., Dalton, Einstein, Millikan, and others). Participant #2 referred to the "Baconian inductive ascent" in the context of this item, which shows the creative use of a term introduced by Lakatos (1970) and discussed in class during a previous course.

While discussing Bohr's model of the atom, Lakatos (1970) concluded: "Bohr's problem was not to explain Balmer and Paschen series, but to explain the paradoxical stability of the Rutherford atom. Moreover, Bohr had not even heard of these formulae before he wrote the first version of his paper" (p. 147). Interestingly, most textbooks consider Bohr's major contribution to be the explanation of the Balmer and Paschen series of the hydrogen line spectrum. Lakatos (1970, p. 147) goes on to show the importance of this event in the history of science as a Baconian inductive ascent: "Since the Balmer and the Paschen series were known before 1913 [year of Bohr's first publication], some historians present the story as an example of a Baconian 'inductive ascent': (1) the chaos of spectrum lines, (2) an empirical law (Balmer), (3) the theoretical explanation (Bohr)." A major premise of historians who follow the Baconian inductive ascent is that scientific theories and laws are primarily driven by experimental observations. Furthermore, such empiricist interpretations consider scientific progress to be dichotomous, that is, experimental observations lead to scientific laws, which later facilitate the elaboration of explanatory theories. In this context, Brush (1978) has explained cogently that:

> as soon as you start to look at how chemical theories developed and how they were related to experiments, you discover that the conventional wisdom about the empirical nature of chemistry is wrong. The history of

> chemistry cannot be used to indoctrinate students in Baconian methods. *(p. 290)*

Hovis and Kragh (1991) have endorsed a similar view: "naïveté of scientists to view scientific development as cumulative validations of theory by distinct, 'closed' experiments—whose data, as 'brute facts,' present no intrinsic interpretative difficulties" (p. 783).

Understanding Scientific Development (Participants' Responses to Item 4)

Item 4 required participants to select one of the three models to understand scientific development as suggested by Windschitl (2004).

Model 1 (Figure 3, p. 505), approximates the traditional linear scientific method characterized by the following steps:

> Observe phenomena → Develop question → Create hypothesis → Design experiment → Conduct experiment → Analyze data → Draw conclusions → New questions arise

Model 2 (Figure 4, p. 506), reflects more of a recursive nature of inquiry based on the conceptual interconnectedness of the phases of inquiry:

> Observe → Develop question ↔ Create hypothesis ↔ Design and conduct investigation ↔ Analyze data → Draw conclusions → Observe

In this model at each stage the scientist can go back and forth in order to facilitate a better understanding of the issues.

Model 3 (Figure 5, p. 507), subsumes the phase interconnection feature of Model 2 and emphasizes that questions should flow from observations framed by a theoretical model and findings are interpreted within the context of how well they support this tentative model:

> Select tentative model → Observe ↔ Develop question ↔ Create hypothesis ↔ Design and conduct investigation ↔ Analyze data → Argue connections between evidence and claim → Argue for revised model → Select tentative model

Given the experience acquired by the participants during class discussions and readings, all of them selected Model 3 (Windschitl, 2004, Figure 5, p. 507). However, what is more important is the conceptualization of scientific development and the following are some of the examples that illustrate participants' thinking:

Model 1 is linear, vertical and explains scientific development by emphasizing experiments (observations). It is a positivist model in which "objective" observations lead to a hypothesis, which in turn suggests experiments that provide data for a theory. In my opinion, this model emphasizes simple observations and ignores the presuppositions. Model 2 approximates more to reality, as it is not so linear and permits retracing steps if the analyses of data do not correspond to expected conclusions. Still I do not accept this model as it emphasizes observations and does not postulate a tentative model. Model 3 is the best as it originates from a tentative model based on theoretical preconceptions of the scientist. This is followed by observations, questions, hypotheses, experiments and analyses of data. The best feature of this model is the possibility of retracing steps based on arguments, until a new model is postulated and ultimately accepted by the scientific community. *(Participant #1)*

An important aspect of Model 3 is its flexibility which allows scientists to rethink and include new considerations during the process of theory construction. Starting from a given model, one can present arguments and then go on to design further investigations. It permits active participation based on experiences that require imagination and creativity. *(Participant #5)*

All observations are based on an implicit theory, which leads to relations, ideas and hypotheses that help to develop a research question. From an epistemological perspective, Model 3 suggests that the role of science is to create, revise and refute theoretical models. This shows the changing or tentative nature of scientific development. This also shows that teaching science should be an investigative process. *(Participant #9)*

Model 1 has a rigid structure and has a vision quite similar to that of the traditional scientific method. Although Models 2 and 3 are quite similar, an important difference is that Model 3 starts with the "selection of a tentative model" whereas Model 2 starts with observations. Furthermore, Model 3 does not require the drawing of isolated conclusions, but rather attempts to connect the data with the initial tentative idea. These two differences make Model 3 more acceptable and the postulation of an initial tentative model is comparable to the "heuristic principle" or "presuppositions" of a scientist before starting an investigation. The search for connections between the data and the heuristic principles is an important aspect of the nature of science. Precisely, based on different presuppositions the same data can also be interpreted differently, leading to rivalries and controversies. This shows yet another facet of how science in the making is also self-correcting. *(Participant #11)*

> Model 3 shows that arguments are an explicit component of scientific research and that knowledge is tentative, which shows the need for continual critical appraisals. The cyclic character of the model makes it very flexible so that we do not necessarily have to finish one phase before starting another one. *(Participant #16)*

Most of the participants not only agreed with Model 3 but also drew attention towards the drawbacks involved in Models 1 and 2. Model 1 was considered to be rigid and represented the traditional scientific method based on a positivist epistemology. Model 2 was considered to approximate more to reality and permitted retracing different phases of the investigative process. However, it was considered to be too empirically oriented. Model 3 was considered to represent scientific development closely, and participants considered the following aspects important:

a. Originates from a tentative model based on theoretical preconceptions of the scientist.
b. Possibility of retracing steps based on arguments.
c. Flexibility in the construction of new models.
d. Active participation based on experiences that require imagination and creativity.
e. The role of science is to create, revise, and refute theoretical models.
f. The tentative nature of scientific development.
g. Connects the data with the initial tentative idea.
h. Interpretation of data based on the heuristic principle or presuppositions of the scientist.
i. Based on different presuppositions the same data can also be interpreted differently, leading to rivalries and controversies.
j. How 'science in the making' is also self-correcting.
k. The need for continual critical appraisals.

All of these aspects are not necessarily based on Windschitl (2004). It is important to note that this item formed part of the Final exam in the last session of the course. This means that teachers had participated in 36 hours of classroom presentations and discussions and thus acquired considerable experience with respect to the issues involved. According to Windschitl (2004):

> Most teachers use facts and concepts as their objects of instruction. These are more meaningful, however, when taught within the contexts of models. Models emphasize causal, functional, temporal, spatial, developmental, and conceptual interrelationships that give meaning to otherwise noncontextualized facts and concepts. The theoretical character of some models also is an entry into discussion of the social construction of science ideas and the tentative nature of scientific knowledge. *(p. 506)*

Indeed, participants in this study went beyond the stage of understanding scientific development as reporting of facts and concepts, namely "a Baconian gathering of facts about the world" (Chinn & Malhotra, 2002, p. 187). Most participants understood the conceptual complexity of the investigative process that facilitates the elaboration of a series of models based on explanatory frameworks. It is precisely this continual critical appraisal that leads to the tentative nature of scientific knowledge (Niaz et al., 2003).

Additional Research Questions

Besides the four evaluation items that formed part of the previous sections of this study, participating teachers were also asked to respond to the following two-item questionnaire:

Q1: Do you think there is a science that we do not teach? (as suggested by Campanario, 1999, Reading 1). Please explain.

Q2: According to Delval (2002, p. 8): "If we do not work for teacher training and change, then all efforts to change education in our schools would be sterile" (cited in Mellado, 2003, p. 352, Reading 2). Please provide a description of some courses that you did not take as an undergraduate that could have helped in your teaching profession.

These two items were provided to the participants at the end of the third day of the final week of the course and they were allowed to take home and submit their written responses the next day.

There is a Science that We Do Not Teach (Participants' Responses to Item Q1)

All participants agreed that there is a science that we do not teach and the following responses illustrate participants' thinking:

> Of course there is a science that we do not teach, based on the premise that we teach science content and not "nature of science," (NOS), which makes understanding science more difficult. Issues related to NOS need to be debated in the classroom if we want our students to understand 'science in the making.' For example it would be very helpful and motivating for students if the following issues are discussed: Where does an original idea come from?, Why do scientists have to publish their work?, Is competition between scientists inevitable?, Who finances research?, and How to recognize the merit of a scientist? These issues can perhaps be included at the end of a chapter in sections such as: Novel aspects of science, New tendencies in research and education, Science for the future, etc. *(Participant #1)*

Science has been considered to be a rigid body of preexisting knowledge and the role of the educator and of the curricula has been simplified and consists in the assimilation and later transmission of knowledge. It is because of this that there exists a science that we do not teach. Campanario (1999) refers to this as metascientific aspects of science and has emphasized that these be included in the classroom as early as primary education. What makes things difficult is that even we as teachers do not have a clear understanding of these NOS related aspects. *(Participant #7)*

The science that we teach is basically empirical science, based on the scientific method and devoid of all controversies and difficulties. This leaves out the heuristic aspects that is, the capacity to invent, interpret and understand experimental data. History of science shows that empirical data in itself does not contribute towards scientific development. Interpretation of alpha particle experiments by J.J. Thomson and E. Rutherford is a good example of how the same experimental data led to two different atomic models [this was discussed in a previous course, cf. Niaz, 1998]. Our textbooks by following the scientific method and the pursuit of "truth" make things even more difficult. *(Participant #8)*

The science that we teach is based on the dominant paradigm which emphasizes the empiricist epistemology. This, consequently, leads to a science that we do not teach. Campanario (1999) considers this as an "immaculate activity" conducted by disinterested scientists who look for "truth" based on a method that can differentiate between science and non–science.Smith and Scharmann (1999), although they do espouse the inclusion of NOS, still restrict some topics for secondary school science based on the criteria of science and non–science.An example from the history of science illustrates this issue. Besides J.J. Thomson two German scientists (Kaufmann & Wiechert) also determined the charge to mass ratio of cathode rays at the beginning of the 20th century [this was discussed in a previous course, cf. Niaz, 1998]. However, it was Thomson who based on his empirical data postulated an atomic model that was accepted by the scientific community. Does this mean that Thomson's work was scientific and that of the German scientists was non–scientific?*(Participant #11)*

There is a science that we do not teach. For example, we present to our students Millikan's oil drop experiment which helped him to determine the elementary electrical charge [cf. Holton, 1978]. However, neither the textbooks nor we tell our students that Millikan discarded oil drops that in many cases did not support the result he was expecting. *(Participant #12)*

It seems that in order to grasp the essence of the NOS, it is important to discuss issues such as: (a) Where does an original idea come from? (b) Is competition between scientists inevitable? (c) How do we recognize the merit of a scientist? It is plausible to suggest that the experience provided in this course led participating teachers to engage in such thought-provokingideas (see Chapter 1 for similar questions that were considered to constitute the NOS). These responses clearly show how and why the way we teach science leaves out important aspects of the NOS, and this leads to a science that we do not teach (Campanario, 1999). Interestingly, teachers not only agreed with the affirmation but also provided examples from the history of science that would constitute the science that we do not teach, such as: The Thomson–Rutherford controversy in the interpretation of alpha particle experiments; The determination of the charge to mass ratio of cathode rays by Thomson on the one hand and Kaufmann and Wiechert on the other differed precisely in their interpretation; and the Millikan–Ehrenhaft controversy with respect to the determination of the elementary electrical charge. It is important to note that these controversial historical episodes (and others) were discussed in a previous course and participants found them to be relevant in the present context, which shows creative adaptation. Campanario (2002) has shown how both students and scientists offer resistance to new ideas and that the inclusion of controversial historical episodes (the metacognitive dimension of science) can stimulate students' intellectual curiosity. With respect to the differentiation of science and non-science (Participant #11), it is important to take into consideration Smith & Scharmann's (1999) advice with respect to the complexity of some NOS issues, especially for secondary school students (also see Scharmann & Smith, 2001). This requires the selection of appropriate materials related to issues of the NOS (science that we do not teach) that are generally left out by both the textbooks and the traditional science curriculum.

Courses that Could Help Teachers to Understand the Nature of Science (Participants' Responses to Item Q2)

Participants argued convincingly for the inclusion of the following courses as part of the undergraduate curriculum in order to understand scientific development: (a) The history, philosophy, and epistemology of science; (b) The nature of science and its implications for education and society; (c) Didactics of science; (d) Epistemology of the teaching of science, Methodology of investigation; (e) Psychological and learning theories and constructivism; (f) Research in the teaching of chemistry. The following responses illustrate participants' thinking:

> To be honest my initial vocation was to be a scientist and not a teacher. However, as a chemist I received a lot of information about chemistry content but none about how chemists work. Again, when I became a teacher

> I received information about education, but very little with respect to how and what to teach. I am concerned that many teachers are still receiving the following advice based on "ancestral" ideas such as: "If you strictly follow the scientific method, as did Newton, Dalton, Einstein, Bohr and others, then you will be doing science and one day will become a scientist like them." *(Participant #1)*

> As teachers we must know that there is no "absolute truth" and that we live in a world in which changes in our knowledge are progressive. This means that we do not have to be constrained by particular theories, models and teaching strategies. However, the reality is different in which the university assumes that it is enough to have content knowledge of a subject, and this precisely is problematic for somebody starting a career in teaching. *(Participant #2)*

> In the 9th grade chemistry program there is no mention of the nature of science (NOS). Inclusion of NOS would help to change the positivist ideas that knowledge is accumulative, lineal and free of controversies. Instead, we need to emphasize the heuristic principles. *(Participant #9)*

> From the very beginning I would have liked to learn, "where did the scientific laws come from," that is, their historical context. This would have provided me a better perspective and even perhaps an incentive to become a research scientist. On the contrary, we were taught that scientists are geniuses and especially "equipped" for doing science. Again, in our undergraduate methodology course we were asked to strictly follow the scientific method. *(Participant #11)*

> I am a graduate of a pedagogical school and we saw two methodology courses at the undergraduate level. However, the methodology course at the graduate level was so different that it changed my vision of science and its development. What we really need is the historical background of the development of science, based on controversies and rivalries among scientists. In other words science in the making is not a bed of roses. *(Participant #13)*

The idea that the success of famous scientists depends on them following the scientific method strictly ("ancestral ideas," as one participant expressed it), is quite widespread in many parts of the world. Both the science textbooks and the traditional curricula endorse this perspective which makes any change even more difficult (cf. Niaz, 2008b). Studies conducted by historians and philosophers of science over the last 50 years, have changed the historical landscape completely. Most philosophers of science would agree that famous scientists such as, Newton,

Dalton, Darwin, Einstein, Bohr, and others did not follow the scientific method as espoused in methodology courses and science textbooks (Cartwright, 1983, 1989, 1999; Giere, 1999, 2006; Holton, 1978, 1998; Kuhn, 1970; Lakatos, 1970; Polanyi, 1964). Phillips (2005), a leading educational theorist has cautioned that in the light of the new history and philosophy of science scholarship even some of the reform documents (NRC, 2002) may not consider the work of famous scientists as rigorous research. Intellectual curiosity of students at an early stage in their careers can easily be converted into an asset by exploring and discussing issues such as: Where do the scientific laws come from? Progressive changes in our scientific theories show the tentative nature of science and that "absolute truth" is always elusive; Scientists are creative and generally do not follow the scientific method; and Scientists interpret data based on their heuristic principles (guiding assumptions, cf. Laudan, Laudan, & Donovan, 1988), which inevitably leads to controversies and rivalries.

Reliability of Results Reported in this Study

In order to provide reliability of results reported in this section (Items 1, 2, 3, 4, Q1, Q2), participants' interventions during the question–answer sessions in the 13 class presentations were checked (that included written questions and comments; see Multiple Data Sources in the previous section for the pool of 250 questions and comments). The objective of checking was to compare participants' responses on an issue on three different opportunities in order to see if they expressed the same underlying idea (triangulation of data sources). First, participants' responses on the written exam question were checked. Second, whether a similar question/comment was expressed twice on a related issue during the question–answer sessions in the 13 class presentations. In the case of some uncertainty or ambiguity, the instructor's class–notesand the critical essay were also consulted. A similar procedure was followed in analyzing participants' responses to all items. This procedure is based on the theoretical framework presented in the section on Multiple Data Sources.

Conclusion: 'Science in the Making' is Not a Bed of Roses

Based on classroom discussions, participants in this study acquired considerable experience with respect to the NOS within an HPS perspective. Exam questions in the study were designed in order to progressively facilitate greater understanding of NOS–relatedissues. Item 1a provided an opportunity to recognize the complexity of the issues involved in 'science in the making.' Item 1b asked participants to select four aspects of the NOS that they considered most important and the following aspects were selected: (1) The tentative nature of scientific theories ($n = 13$); (2) Observations are theory laden ($n = 9$); (3) The role of conflicts, controversies, and rival theories in scientific progress ($n = 8$); (4) Scientists use their imagination and creativity to develop theories (n = 7); (5) Scientific knowledge relies on experimental evidence, rational arguments, and skepticism (n = 6); (6) Objectivity in science is a consequence of a social process (n = 5); (7) There is no universal scientific method (n = 4); and (8) Science is self-correcting(n = 4). These are important aspects of the NOS and have been recognized by researchers in science education (McComas, 2008). Item 2a provided an opportunity to consider the difficulties involved in teaching the NOS and participants considered the new vision to contradict some of the objectives of the traditional science curriculum based on positivism. Item 2b asked participants to select one aspect of the NOS that in their opinion was most important and the following were selected: (i) The tentative nature of scientific theories (n = 7); (ii) The role of heuristic principles: observations are theory laden (n = 5); and (iii) Competition among rival theories: the same experimental data can be interpreted differently (n = 4). The rationale behind this item was to go beyond Item 1b in which participants selected four important aspects of the NOS. Interestingly, in Item 1b participants referred to eight different aspects of the NOS and this number reduced to three on Item 2b. The tentative nature of scientific theories was considered to be the most important aspect of the NOS in both Item 1b (selected by 13 participants) and Item 2b (selected by seven participants), and all seven participants who selected this aspect on Item 2b also did so on Item 1b. Five participants manifested a creative adaptation by considering the role of heuristic principles as that of guiding assumptions (the theory-ladennature of observations).

All participants except one disagreed with the affirmation: "Scientific theories emerge in a logical manner" (Item 3a). Participants reasoned cogently by referring to the theory-ladennature of observations in the context of different historical episodes. One participant reasoned that the oil drop experiment could not have been performed (design) without Millikan's guiding assumption. Based on Item 3b, a majority of the participants (n = 12) understood the dual role of a hypothesis in foreseeing the results and providing a tentative explanation and that elaboration of a hypothesis is, in itself, a complex process based on presuppositions and heuristic principles. All participants understood that the most important aspect in the development of a theory is the heuristic principle and not the experimental data (Item 3c). Most historians and philosophers of science would not recommend "Baconian methods" in which scientific theories and laws are primarily driven by experimental observations (Brush, 1978; Hovis & Kragh, 1991; Lakatos, 1970).

All participants understood that reporting of facts and concepts in scientific development is important in the context of models (Item 4). Elaboration of models requires considerable conceptual complexity based on: theoretical preconceptions of the scientist, possibility of retracing steps, flexibility in the construction of models, interpretation of data based on heuristic principles, imagination, and creativity.

The additional research question Q1, required participants to consider the following controversial aspect, "Do you think there is a science that we do not teach?" All participants agreed based on the premise that we teach science content and not the NOS. Participants not only agreed that the way we teach science leaves out important aspects of scientific development but also provided examples from the history of science that would constitute the science that we do not teach. Based on additional research question Q2, participants were critical of their undergraduate science education in which both the science curricula and textbooks endorse the view that the success of famous scientists depends primarily on them following the scientific method strictly.

One of the most important aspect of participants' responses to all items in this study was that they tried to substantiate their arguments based on different episodes from the history of science that formed part of this and previous graduate courses. This postpositivist landscape based on historical reconstructions was different from their undergraduate courses based on an empiricist epistemology and often led to conflicting situations (cf. Mellado et al., 2006 for further elaboration). Finally, given the conceptual complexity of the scientific endeavor it is plausible to suggest that 'science in the making' is not a bed of roses.

5

THE ROLE OF HEURISTIC PRINCIPLES IN UNDERSTANDING THE NATURE OF SCIENCE

Introduction

According to Schwab (1974) scientific inquiry tends to look for patterns of change and relationships which constitute the heuristic (explanatory) principles of our knowledge. In other words:

> A fresh line of scientific research has its origins not in objective facts alone, but in a conception, a deliberate construction of the mind. On this conception, all else depends. It [heuristic principle] tells us what facts to look for in the research. It tells us what meaning to assign these facts. *(Schwab, 1974, p. 164)*

Monk and Osborne (1997) pointed out how many science curricula have forgotten Schwab's important epistemological distinction between the methodological (experimental data) and interpretative (heuristic principles) components. Matthews (1994a) emphasized the importance of heuristic principles in scientific inquiry and science education in similar terms. To understand the function of "heuristic principles" it is helpful to consider J.J. Thomson's (1897) experimental work with cathode rays, which helped to elucidate the structure of the atom (see Chapter 2 for details). Although the experimental details are important we cannot ignore the rationale behind Thomson's determination of the charge to mass ratio of cathode rays.

The history of the structure of the atom since the late nineteenth and early twentieth centuries shows that the models of J.J. Thomson, E. Rutherford, and N. Bohr evolved in quick succession and had to contend with competing models based on rival research programs. This period of the history of the structure of the atom has been the subject of considerable debate and controversy in the history and philosophy of science (HPS) literature (Achinstein, 1991; Falconer, 1987; Heilbron, 1981; Heilbron & Kuhn, 1969; Hettema, 1995; Holton, 1988, 1993; Jammer, 1966; Kuhn, 1984; Lakatos, 1970; Popper, 1965). Blanco and Niaz (1997) have shown how both students and teachers understand Thomson's experiments as a series of conclusions (a rhetoric, according to Schwab) based on empirical findings (truths). In the case of Bohr's research program, Lakatos (1970) considers Bohr's explanation of the paradoxical stability of the Rutherford atom as the heuristic principle. In contrast, most textbooks consider Bohr's major contribution to be the explanation of the Balmer and Paschen series of the hydrogen line spectrum (i.e., experimental findings). This reminds us that for almost 50 years we have ignored Schwab's (1962) advice that science cannot be taught as an "unmitigated *rhetoric of conclusions* in which the current and temporary constructions of scientific knowledge are conveyed as empirical, literal, and irrevocable truths" (p. 24, emphasis in original).

According to Schwab (1974), it is important to understand not only the experimental details but also the heuristic principle that underlies the experimental findings:

> In physics, similarly, we did not know from the beginning that the properties of particles of matter are fundamental and determine the behavior of these particles, their relations to one another. *It was not verified knowledge but a heuristic principle, needed to structure inquiry, that led us to investigate mass and charge and, later, spin. (Schwab, 1974, p. 165, emphasis added)*

The nature of science (NOS) has been the subject of considerable research in science education (Abd-El-Khalick2005; Akerson, Morrison, & Roth McDuffie, 2006; Blanco & Niaz, 1997; Cobern, Gibson, & Underwood, 1999; Dogan & Abd-El-Khalick,2008; Khishfe & Lederman, 2006, 2007; Lederman, Abd-El-Khalick, Bell, & Schwartz, 2002; Niaz, 2001a, 2008b, 2009a; Osborne, Collins, Ratcliffe, Millar, & Duschl, 2003; Smith & Scharmann, 1999, 2008). Despite some differences, a fair amount of consensus has been achieved in the science education community with respect to the following aspects of the NOS (Niaz & Maza, 2011):

1. Scientific knowledge relies heavily, but not entirely, on observation, experimental evidence, rational arguments, and skepticism.
2. The theory-ladennature of scientific observations (role of guiding assumptions).
3. The tentative and fallible nature of scientific knowledge.
4. There is no one way to do science and hence no universal step-by-stepscientific method can be followed.
5. Laws and theories serve different roles in science and hence theories do not become laws even with additional evidence.

6. Scientific progress is characterized by competition among rival theories.
7. Different scientists can interpret the same experimental data in more than one way.
8. Development of scientific theories at times is based on inconsistent foundations.
9. Scientists require accurate record keeping, peer review, and replicability.
10. Scientists are creative and often resort to imagination and speculation.
11. Scientific ideas are affected by their social and historical milieu.

Based on these considerations the objective of this study is to facilitate chemistry teachers' understanding of the role played by heuristic principles in understanding the NOS.

Rationale and Design of the Study

This study is based on 26 in–servicechemistry teachers who had enrolled in the course, "Investigation in the Teaching of Chemistry" as part of a Master's degree program in education at a major university in Latin America. Ten teachers worked in secondary schools and 16 at the university level (female = 16, male = 10, age range: 25–40 years), and their teaching experience varied from 5 to 15 years. In the previous year all teachers had enrolled in the course, "Methodology of Investigation in Education," in which the basic philosophical ideas of Popper, Kuhn, and Lakatos were discussed in order to provide an overview of the controversial nature of progress in science and its implications for research methodology in education. Teachers were aware that basic ideas like the scientific method, objectivity, and the empirical nature of science were considered to be questionable by philosophers of science.

Course Content (Reading List)

The course was based on 18 required readings and was subdivided into the following units:

- *Unit 1: History and philosophy of science in the context of the development of chemistry*: Readings: (1) Matthews (1994b); (2) Campanario (2004); (3) Niaz (1998); (4) Niaz (2000b); (5) Chamizo & Gutiérrez (2004); (6) Niaz (2001c); (7) Brito, Rodríguez, & Niaz (2005).
- *Unit 2: Students' alternative conceptions*: Readings: (8) Furió, Azcona, & Guisasola (2002); (9) Kousathana, Demerouti, & Tsaparlis (2005); (10) de Posada (1999); (11) Niaz (2000c); (12) Campanario & Otero (2000).
- *Unit 3: Conceptual change in learning chemistry*: Readings: (13) Niaz (1995b); (14) Niaz et al. (2002); (15) Niaz (2002); (16) García (2000); (17) Marín, Solano, & Jiménez (1999); (18) Niaz (2005b).

Course Organization and Activities

On the first day of class (two hours) all participants were provided copies of all the readings and salient features of the course were discussed. It was emphasized that the course called for active participation. As all teachers worked in nearby schools and universities, three types of course activities were programmed:

a. Class discussions were planned on Saturday of the 5th and 7th weeks of the course (three hours in the morning and three in the afternoon). Readings 1–9 were discussed in the first meeting and readings 10–18 in the second meeting. Teachers were supposed to have studied each of the readings before the meetings. Each meeting started off with various questions and comments by the participants. The instructor intervened to facilitate understanding of the issues involved. Throughout the 10 weeks of the course, the participants communicated among themselves and the instructor through emails (total time devoted to class discussions = 12 hours).
b. Class presentations by the participants were programmed during the 10th and the final weeks of the course (Monday to Saturday, total time = 44 hours). On the first day of class all participants (some in groups) selected (by a draw) one of the 18 readings for a presentation. Each participant or group was assigned 90 minutes (30 minutes for the presentation and 60 minutes for interventions and discussions). Each of the presentations was moderated by one of the participants. The instructor intervened when a deadlock was reached on an issue. It was expected that the participants would present the important aspects of the readings, with the objective of generating critical discussions. All participants prepared PowerPoint presentations.
c. During class presentations in the final (10th) week participants were encouraged to ask their questions in writing, which were then read out loud by the moderator. The presenters were given the opportunity to respond and then a general discussion followed. At the end of each session all written questions were submitted to the instructor, which provided important feedback with respect to issues, conflicts, and interests of the participants. Each participant signed her/his question. The same procedure was followed in all 18 presentations, generating a considerable amount of data (a pool of 510 questions and comments). Besides this data, the instructor also took class–notesthroughout the course. These data facilitated the corroboration (triangulation of data sources) of participants' responses to the three exam questions in this study.

Course Evaluation

All participants presented an Initial exam in the first session and a Final exam during the last session of the final week (10th). The rationale behind these evaluations in this study (could have been named differently) was not merely evaluation in the traditional sense. Both exams were "open book" (about three hours each) and participants were allowed to consult any material that they felt could be helpful. The Initial exam reflected participants' experience during the first nine weeks, including 12 hours of class discussions. The Final exam reflected an additional experience of 36 hours of class presentations and discussions during the final week. Furthermore, both exams were based on the premise that oral presentations and discussions can be complemented by written responses and thus facilitate greater understanding. This study is based on participants' written responses to two items of the Initial exam and one item of the Final exam. These three items were formulated in order to provide possible answers to the objective of this study. The three items included in the Initial and Final exams are presented below:

- *Initial exam*
  - *Item 1*:
    - (a) Select five aspects of this course that in your opinion are novel.
    - (b) How can these aspects change your teaching practice in the classroom?
  - *Item 2*: Given the importance of the different aspects of the "nature of science" (Reading 18, Niaz, 2005b, p. 411), in your opinion: What are the factors that impede the implementation of new strategies in teaching chemistry?
- *Final exam*
  - *Item 3*: According to Schwab (see Readings 1, 3, 4, 14, and 18) in the development of science there exists a very important difference between methodological aspects (empirical data) and the interpretations (heuristic principles). These interpretations inevitably generate controversies and polemics. Similarly, Schwab has emphasized that textbooks resort to "rhetoric of conclusions" instead of convincing students with arguments. Nevertheless, research has shown (see Reading 14) that students often do not accept the arguments (new ideas) easily and offer resistance to conceptual change. How would you resolve this dilemma in your classroom?

Multiple data sources

Based on the different course activities this study generated the following data sources:

a. Question–answer sessions after each of the 18 formal presentations, in which participants wrote their questions/comments (interventions), which were then discussed in class and later submitted to the instructor (a pool of 510 written questions and comments were generated).

b. Initial and Final exams during the final week, separated by 36 hours of class presentations and discussions (total time = 8 hours).
c. The instructor's class–notes,based on the following activities throughout the course: class discussions during the 5th and 7th weeks of the course (total time = 12 hours), 18 formal presentations during the final week, question–answer sessions after each presentation. These notes consisted of quotes from the students' (including their names) interventions during presentations and discussions, accompanied by general comments.
d. Critical essay. As part of their evaluation all participants were required to submit a critical essay based on any one or various readings. The objective of this essay was to present a critique based on epistemological, philosophical, and methodological aspects. This essay was submitted 10 days after the Final exam, which provided the participants ample time to reflect and elaborate their ideas.

At this stage it is important to emphasize the role played by triangulation in educational research through multiple data sources. This study has a hybrid research design, in which data were collected through systematic observation (for details see Chapter 4 and Niaz, 2008d). Research reported here is based on actual classroom practice and participants did not feel constrained by the research design. All the data generated (56 hours of class activities) were based on regular classroom activities. An example of a similar study is provided by Niaz (2008b).

Results and Discussion

In this section participants' responses to the three evaluation items are presented. For each of the items all responses were analyzed and, based on this, various categories were generated. Each of the responses was classified into one of the categories. Some of the responses could easily have been classified into more than one category.

Participants' Responses to Item 1a (Select Five Aspects of this Course that in your Opinion are Novel)

Based on the participants' responses, the categories shown in Table 5.1 were generated.

1. The Role of the History and Philosophy of Science in Science Education (n = 24)

The following responses illustrate participants' thinking:

> Honestly, most of the aspects related to HPS [History and philosophy of science] were a surprise to me, as it was in sharp contrast to my prior

TABLE 5.1 Participants' Responses to Item 1a (Select Five Aspects of this Course that in your Opinion are Novel)

*Response Category**	*Frequency (n)*
1. The role of the history and philosophy of science in science	24
2. The importance of the nature of science	24
3. Conceptual change in learning science	18
4. Students' alternative conceptions	13
5. Critical and reflective thinking	13
6. Research articles for questioning inadequate ideas of teaching science	13
7. Motivation for doing further research	10
8. Heuristic principles and rhetoric of conclusions	9
9. Science is objective only in a certain context of scientific development	9
10. Cognitive psychology for meaningful learning	5

* These categories were generated based on participants' responses to Item 1a.

> conceptions of science. I must point out that I was trained as a chemical engineer. Despite efforts to the contrary, I still emphasize the empirical nature of science in my classes. I am now trying to reorient my classes in order to present to the students appropriate aspects of the history of science. *(Participant #10)*

> Inclusion of the history and epistemology of science is all the more important as very little of this aspect is presented in our textbooks. From this perspective it is much more motivating to understand science, and especially that it is not a simple "rhetoric of conclusions"' *[Schwab, 1974]. (Participant #11)*

> Some of the experiments included in the course have formed part of our scientific career for many years, and still we were not aware of the historical background that contributed to the development of theories, laws and principles that we use so frequently for the resolution of our problems. Furthermore, many of these experiments led to controversies among the scientists that lasted for long periods, as it was difficult to interpret the results. *(Participant #20)*

These responses clearly show the importance of HPS in order to understand the historical background of the laws and theories that are discussed so frequently in the chemistry curriculum. It is precisely this lack of a historical perspective that leads the textbooks and the teachers to emphasize the conclusions as rhetoric, which are presented as algorithms and inevitably lead to their memorization by the students.

2. The Importance of the Nature of Science (n = 24)

The following responses illustrate participants' thinking:

> Atomic models of J.J. Thomson and E. Rutherford are well known to us. However, I was not aware that in contrast to Rutherford's hypothesis of single scattering, Thomson had put forward the hypothesis of compound scattering (for the same data from alpha particle experiments), in order to defend his "plum–pudding"model of the atom. In the case of the periodic table, I was not aware of the controversy with respect to the placing of argon [Reading 7]. In this case Mendeleev's role was surprising indeed: on the one hand he defended his perspective and on the other was open to criticisms. *(Participant #1)*

> Precisely, teaching within science facilitates understanding of the "nature of science." *(Participant #3)*

> Science changes continually and in the process enriches our knowledge. In contrast, some people think that everything has already been found which constitutes an absolute truth. It is important to note that experiments are important, but their interpretation is even more important. *(Participant #7)*

> Controversies are an important part of science. In the case of the Millikan–Ehrenhaft controversy, two scientists interpreted the same experimental results differently. Millikan was able to convince the scientific community and won the Nobel Prize. Later, however, it was found that he did not include all the experimental data [Holton, 1978]. Surprisingly, at present (based on textbooks) we teach our students that Millikan used an exact and beautiful procedure with no reference to the controversy. *(Participant #17)*

These participants considered the NOS to be an integral part of science. In this context, the role of controversies was explicitly recognized (Millikan–Ehrenhaft, Thomson–Rutherford, Mendeleev and the placing of argon). Furthermore, it is important to note that teachers pointed out how their professional training in science did not provide the background that could have enriched their knowledge.

3. Conceptual Change in Learning Science (n = 18)

The following responses illustrate participants' thinking:

> What I found interesting and novel was that research has not only found students' alternative conceptions but also designed possible teaching strategies in order to facilitate conceptual change. *(Participant #11)*

> Previously, I was not aware of the difference between algorithmic and conceptual problems [Reading 13] and considered numerical problems found in most textbooks to be very important. In order to facilitate the transition from algorithmic to conceptual we need to "teach in order to understand." *(Participant #17)*

> In order to understand chemistry it is not sufficient to memorize a formula (found in textbooks anyway) and solve algorithmic problems. As this course has shown, in order to go beyond this paradigm we need to facilitate conceptual change by incorporating theoretical, historical and philosophical aspects. *(Participant #24)*

Most participants (especially those who had an engineering background) agreed that based on textbooks they used to emphasize algorithmic problems based on memorization of content. In the context of the HPS perspective the need to introduce conceptual change is all the more necessary.

4. Students' Alternative Conceptions (n = 13)

The following response illustrates participants' thinking:

> Alternative conceptions enter into conflict with scientific knowledge and frequently convert into what Lakatos (1970) has referred to as the "hard-core" of thinking. Lakatos assumes that the hard-coreof a research program is irrefutable and defended by a series of hypotheses that constitute the protective belt. As the evidence against a research program accumulates it can ultimately be replaced. By analogy, students' alternative conceptions constitute the hard-core and the explanations that provide the justifications, the protective belt. Given the opportunity to question the hard-core of their thinking, students may adopt ideas more in accord with scientific knowledge. *(Participant #8)*

This response (five other participants gave very similar responses) provides an example of how participating teachers can experience conceptual change by transferring knowledge from one context to another. Reading 11 provided teachers an example of how even after having responded correctly in one context that approximates the kinetic view of heat energy students can fall back on the caloric theory of heat in a different context. Based on this it was suggested that an epistemological belief in the caloric theory of heat forms part of the hard-core of students' framework and that the conceptual shift (change) to the kinetic theory requires radical restructuring (Chinn & Brewer, 1993). Participants in this category considered students' alternative conceptions (in general) a part of the hard-coreof beliefs and consequently conceptual change would require strategies that question the hard–core.This clearly provides insight beyond what the course material (Reading 11) had discussed.

5. Critical and Reflective Thinking (n = 13)

The following responses illustrate participants' thinking:

> When students are provided an opportunity to critique their own or other's thinking, it is possible to generate a new perspective, similar to what Lakatos has referred to as competing models or programs, leading to progressive transformations. This of course goes beyond Popper. *(Participant #6)*

> Promoting critical thinking among students is important, as according to Lakatos history of science has been and should be based on competing research programs. Thus students can go back and forth between their alternative conception and the new conception based on scientific knowledge. *(Participant #7)*

> Inclusion of critical and reflexive activities can facilitate greater student participation and thus facilitate not only conceptual understanding but also a positive attitude towards science. *(Participant #12)*

Once again, in this category participants provided a novel interpretation by suggesting that given the opportunity to critique, the two (original perspective and its criticism) constitute rival or competing models, similar to what Lakatos (1970) has suggested in the history of science.

6. Research Articles for Questioning Inadequate Ideas of Teaching Science (n = 13)

The following responses illustrate participants' thinking:

> I was drawn to this aspect as it reveals the dynamics of science and the construction of scientific knowledge. Science is taught in a repetitive manner. In the classroom we teach the way in which we have been taught and tell our students that what has been written and patented cannot be changed. *(Participant #2)*

> Sincerely, I was not aware that research articles could be appropriately adapted for use in the classroom. This could provide students an opportunity to understand the real dimension of scientific research and that scientists are real persons. *(Participant #10)*

> I was so impressed by this article (Reading 2) that I shared it with some colleagues in my school and later we met to discuss and design new teaching strategies. Some colleagues also wanted copies of other reading material included in this course. *(Participant #15)*

> This article has multiple applications: how is research done, how is knowledge generated, how is "life" within the scientific world, motivation for further research and an alternative source of knowledge distinct from the textbook? *(Participant #16)*

> This course has effectively used the research article as a means to facilitate teachers' understanding of the educational process. Furthermore, most of the articles included were from journals that are indexed in the "Social Science Citation Index," which ensures a high quality of research. *(Participant #22)*

This category of responses is primarily based on Reading 2 which provided an overview of the dynamics of scientific research and how it can be used in the classroom. Most participants were not aware that research based on classroom activities is an important area of study and can easily be adapted for classroom use and thus complement the traditional textbook.

7. Motivation for Doing Further Research (n = 10)

The following responses illustrate participants' thinking:

> Like my colleagues, I used to think that our actual textbooks were adequate and that this was the only way of writing them. However, based on readings in this course, that have analyzed textbooks from a historical perspective, my views have changed. I am now motivated to do further research in order to facilitate conceptual understanding of my students. *(Participant #7)*

> First let me recall: This world does not belong to those who are simply "followers," but instead to those who are "innovators." I am now fully aware that in order to improve our education, we must do research and thus innovate. *(Participant #22)*

Participants in this category not only recognized the importance of research articles for the classroom, but went further by recognizing the need for teachers themselves to do research and innovate.

8. Heuristic Principles and Rhetoric of Conclusions (n = 9)

The following responses illustrate participants' thinking:

> In my opinion, some of the difficulties involved in students' and teachers' conceptual understanding are related to the "heuristic principles" [Schwab, 1974]. It is precisely these underlying principles of which we are generally not aware that cause "noise" when we need to go beyond textbook presentations that constitute, "rhetoric of conclusions." *(Participant #10)*

> Most of us (including the textbooks) consider that science progresses because there is scientific equipment available and a scientist to handle it and consequently arrive at conclusions. This is the standard procedure in which "heuristic principles" are ignored and science is presented as "rhetoric of conclusions" [Schwab, 1974]. This presents to the students a perspective in which other factors, such as affective, sociological and cognitive play no part and science becomes "objective" and theories not subject to change. *(Participant #17)*

These responses clearly show the importance of heuristic principles in understanding the NOS. For most chemistry textbooks and curricula it is the handling of the experimental apparatus that facilitates success and makes science "objective." For example, in the determination of the elementary electrical charge, besides his apparatus Millikan was guided by a heuristic principle, namely the existence of a universal charged particle and the magnitude of the charge (Holton, 1978; Niaz, 2000b, 2009a). Holton (1969) has referred to this as the "experimenticist fallacy" frequently found in science curricula and textbooks. Campbell, a methodologist, has called attention to the pitfalls involved in "experimenticism" in cogent terms:

> The objectivity of physical science does not come from turning over the running of experiments to people who could not care less about the outcome, nor from having a separate staff to read the meters. It comes from a social process that can be called competitive cross-validation. . . and from the fact that there are many independent decision makers capable of rerunning an experiment, at least in a theoretically essential form. *(Campbell, 1988b, p. 324)*

"Competitive cross-validation" is a social process based on peer review and not on the competence of a single experimenter. In other words, "objectivity" in science is the consequence of a social process and not based directly on the experimental procedures used in the lab.

9. Science is Objective Only in a Certain Context of Scientific Development (n = 9)

The following responses illustrate participants' thinking:

> I have "re–discovered"that observations need to be interpreted with care as these are always theory–laden.This framework is very common among students who believe that, "Scientific knowledge is objective." We need to emphasize that experimental data need to be interpreted in order to facilitate conceptual understanding. *(Participant #10)*

> Based on my empiricist formation, the "objective" nature of science provided me with the framework to understand that "Knowledge is out there, and it does not matter how it is presented." Now that I have come to see beyond the positivist view of science, teaching science is quite a challenge and much more interesting. Previously, based on the positivist paradigm, I considered that the ethics of investigation are based entirely on: objectivity, rigorous and straight forward reporting of results. Now I understand that data that support our theories need interpretations. *(Participant #16)*

These responses refer to the theory–ladennature of observations which is considered to be an important characteristic of the NOS. Data by themselves do not constitute scientific knowledge. On the contrary, it is the interpretations of data which inevitably lead to controversies and require competitive cross–validation.It is this social process that facilitates consensus and may be considered as "objective knowledge." Nevertheless, it is important to consider the following advice from Giere (2006):

> the strongest claims a scientist can legitimately make are of a qualified, conditional form: "According to this highly confirmed theory (or reliable instrument), the world seems to be roughly such and such." There is no way legitimately to take the further *objectivist step* and declare unconditionally: "This theory (or instrument) provides us with a complete and literally correct picture of the world itself." *(pp. 5–6, italics added)*

10. Cognitive Psychology for Meaningful Learning (n = 5)

The participants emphasized the importance of Piaget, Ausubel, Bruner, and Vygotsky, by suggesting teaching strategies based on cognitive conflicts, equilibration, meaningful learning, and various forms of constructivism.

Participants' Responses to Item 1b (How Can these Aspects Change your Teaching Practice in the Classroom?)

The following responses illustrate participants' thinking:

> While teaching the different topics I can show to the students that there is no absolute truth and our knowledge changes over time. In other words this can facilitate an understanding of the tentative nature of scientific knowledge. *(Participant #2)*

> I found the strategy of creating problematic situations in the classroom leading to cognitive conflicts, very appealing. This can lead to a reorganization of cognitive structures and later facilitate conceptual change. *(Participant #7)*

> Research published in indexed journals can influence the educational process. This aspect seemed so novel to me that I distributed the article by Campanario (Reading 2) among my colleagues in the school. Most of them were surprised and found it very useful. Based on the readings I came to realize the importance of research and that I too could investigate and publish. *(Participant #9)*

> This semester in my 9th grade class, while teaching atomic structure, instead of including many experimental details, I introduced some historical description of the controversy between J.J. Thomson and E. Rutherford [Reading 3]. This was so motivating for the students that in the next experiment they wanted to know the origin of the experiments and if there were contradictions between the scientists and how were these resolved. *(Participant #15)*

> Most of the teachers depend on textbooks in order to facilitate information to our students. However, as most textbooks distort the content we need to find alternative sources. One way of doing that is looking for research literature based on different concepts and perspectives. *(Participant #25)*

These are interesting responses and provide some evidence for how the teachers considered that their role in the classroom could be more effective if they introduced novel strategies based on: cognitive conflicts, research articles, and historical reconstructions. This, of course, requires that the teacher must necessarily conceive his role as that of a practicing researcher. Based on the experience of these teachers it is plausible to suggest that students find the historical reconstructions interesting and look forward to similar treatments in other experiments included in the curriculum.

Note: In order to provide reliability of results (triangulation of data sources), participants' critical essays and interventions during the 18 class presentations were checked (see section Multiple data sources, for a pool of 510 written questions and comments). It was found that all participants expressed similar ideas on at least two different occasions. Guba and Lincoln (1989) consider such "cross-checkingspecific data items of a factual nature" (p. 241) as an essential part of triangulation of data sources.

Participants' Responses to Item 2 (Given the Importance of the Different Aspects of the "Nature of Science" (Reading 18, p. 411), in your Opinion: What are the Factors that Impede the Implementation of New Strategies in Teaching Chemistry?)

Before presenting results for this item it is important to present the different aspects of the NOS, as presented in Reading 18 (Niaz, 2005b, p. 411). These are quite similar to those presented in the Introduction section:

1. Scientific theories are tentative.
2. Theories do not become laws even with additional empirical evidence.
3. All observations are theory-laden.
4. Science is objective only in a certain context of scientific development.
5. Objectivity in science is based on a social process of competitive cross-validation through critical peer evaluation.
6. Science is not characterized by its objectivity but, rather, by its progressive character—progressive problemshifts (Lakatos, 1970).
7. Progress in science is characterized by conflicts, competitions, inconsistencies, and controversies among rival theories.
8. Scientists can interpret the same experimental data in more than one way.
9. Most of the scientific laws are irrelevant and at best can be considered as idealizations.
10. There is no universal scientific method based on steps to be followed.

Based on the participants' responses, the categories shown in Table 5.2 were generated.

1. Empiricist Presentations in Chemistry Textbooks that Lack a History and Philosophy of Science Perspective, HPS (n = 19)

The following responses illustrate participants' thinking:

> Textbooks limit themselves to present ideas, concepts and theories in their final elaborated (product) form, which does not involve discussions or doubts. Textbook presentations contradict different aspects of the nature of

TABLE 5.2 Participants' Responses to Item 2 (Given the importance of the different aspects of the 'nature of science', reading 18, p. 411, in your opinion: what are the factors that impede the implementation of new strategies in teaching chemistry?)

Response category★	*Frequency (n)*
1. Empiricist presentations in chemistry textbooks that lack a history and philosophy of science perspective	19
2. Schwab's "rhetoric of conclusions" and lack of "heuristic principles"	18
3. Teachers' and students' epistemological views	16
4. Scientific progress devoid of controversy, interpretation of data, and idealization	14
5. Non–recognitionof the tentative nature of scientific knowledge	11
6. Curricular programs and reforms	8
7. Empirical and not theory–ladennature of science	8
8. Objectivity in science based on experimenticism	7
9. Scientific method	6

★ These categories were generated based on participants' responses to Item 2.

> science as demonstrated by various research articles (e.g., atomic structure, determination of the elementary electrical charge, kinetic theory, covalent bond, periodic table, acids and bases). This research clearly shows that in order to give importance to method and experimentation, textbook authors ignore the conceptualization of the idea that leads a scientist to do experiments and "design" theories. *(Participant #10)*

> In contrast to the HPS perspective, the inductivist vision is rigid and does not contemplate "transgressions" of objectivity, precision, and methodology. This rigidity makes any change in a paradigm difficult, and textbooks continue to repeat the same framework. *(Participant #16)*

> Our textbooks (generally written in the 60s and 70s) and their new editions present more color photographs and images. However, there is no change in the positivist perspective with respect to the absolute nature of theories and laws. *(Participant #18)*

> It seems that it is difficult to change textbooks as it seems that most authors and publishers consider that "why change if a textbook is selling well." *(Participant #26)*

It is important to note that a majority of the participants considered their textbooks to be an important obstacle in implementing new teaching strategies. Participants were particularly aware of the inductivist framework of the textbooks that imposed a vision of science based on science as a finished product (final form, cf. Duschl, 1990) with no attempt to facilitate an understanding of how theories are constructed. Furthermore, it is suggested that scientists do experiments, and elaborate theories with no recognition of the difficulties involved in interpreting data. This, of course, leads to a dilemma: if scientists are absolutely "objective" then the path from data to theory would be free from controversies. The NOS and the historical reconstruction of the different episodes provide a picture in which progress in science is replete with controversies and rivalries.

2. Schwab's "Rhetoric of Conclusions" and Lack of "Heuristic Principles" (n = 18)

The following responses illustrate participants' thinking:

> Teaching chemistry generally means the presentation of "truths" that lack a vision based on history and philosophy of science. Furthermore the different events that led to the formation of scientific knowledge are presented within a perspective that is impregnated with a "rhetoric of conclusions" that ignores the "heuristic principles" (Schwab, 1962). *(Participant #6)*

> Lack of "heuristic principles" lead students and teachers to use consciously or unconsciously a "rhetoric of conclusions," and thus we present the "tentative" as "definitive," without explaining what were the premises, interpretations and evidences that led to a particular conclusion. *(Participant #10)*

> As teachers we generally present the opinion of a famous scientist as authoritative and the absolute "truth." This leads the students to memorize the "truth" and ultimately leads to the traditional "rhetoric" of science. *(Participant #20)*

Once again a majority of the participants were aware of how science is taught with no reference to the guiding assumptions of the scientists, namely, why and how a scientist decides what needs to be researched (cf. heuristic principles, Schwab, 1962). Contrary to all historical evidence, we continue to teach science as a monolithic "rhetoric" that needs no scrutiny or understanding. If a scientist has access to absolute "truth" it only needs to be memorized and not scrutinized and hence we end up with a "rhetoric of conclusions" in the classroom.

3. Teachers' and Students' Epistemological Views (n = 16)

The following responses illustrate participants' thinking:

> As teachers we teach chemistry as we were taught. Our educational background was strictly based on the empiricist tradition, namely it is

> possible to "do science" by strictly following the scientific method. *(Participant #10)*

> We teach as we were taught and it seems that it is difficult to change the "model." *(Participant #16)*

> By ignoring the relevant historical aspects the teacher is "scientifically illiterate." *(Participant #21)*

> In order to introduce changes in our teaching practice we need a new (emergent) paradigm. *(Participant #22)*

Interestingly, teachers in this study considered their own and students' epistemological views as an impediment in introducing new NOS-relatedteaching strategies. This is understandable (despite the reform efforts, AAAS, 1993), as in most parts of the world, science curricula, textbooks, and teachers themselves are deeply imbued by the scientific method and the empiricist tradition. Any change will require a considerable amount of work in rewriting textbooks and teacher training courses and workshops.

4. Scientific Progress Devoid of Controversy, Interpretation of Data, and Idealization (n = 14)

Participants' responses to aspects 7, 8, and 9 of the NOS (cited above, Reading 18, p. 411) are combined in this category and the following examples illustrate their thinking:

> It is difficult to teach with a critical perspective, when controversies, conflicts, presuppositions and idealizations are not only denied but also "concealed" from students. Precisely, this leads to a distortion of the reality. Actually, science is presented as "absolute" and "clean" very much within the positivist paradigm which seems to pervade all aspects of our education. *(Participant #1)*

> Different aspects of the nature of science (including controversies and idealization) constitute the emerging vision. However, in many parts of the world positivism still wields considerable influence, according to which "experiment and observation" are the hallmark of "scientifically proved objective knowledge." *(Participant #11)*

> Science can be conceived as a "sacred affair" and we can establish an analogy between paradigm transitions (based on controversies and rivalries) and crucial moments in religious history—in both transgressions are not

> well received and penalized. Similarly, science teachers reject novel and alternative ideas that contrast with the information provided in textbooks. *(Participant #16)*

Although, scientific controversies, conflicts, presuppositions, and idealizations, are found throughout the history of science, yet classroom practice not only denies these aspects but also tries to conceal them from students. It is important to note that science curricula and textbooks in general do have a philosophical framework (Marquit, 1978), namely an all pervading positivist milieu based on "scientifically proved objective knowledge." Indeed, this is a very pithy argument for introducing an alternative framework based on new ideas that included HPS in the textbooks and eventually in the classroom. Actually, it is quite astonishing as to why we deny students a critical perspective with respect to how scientists go about doing science (Niaz, 2010a, 2011). As one participant suggested, transgressions are not welcome either in the history of science or in religious history, and this perhaps provides insight as to why teachers are also reluctant to accept novel and alternative ideas.

5. Non-recognition of the Tentative Nature of Scientific Knowledge (n = 11)

This refers to aspect 1 of the NOS (Reading 18, p. 411) and the following responses illustrate participants' thinking:

> This is the most difficult aspect. It is very difficult to teach students with this perspective as there is a well–establishedtradition in which it is inculcated that scientific theories are universal and unequivocal. *(Participant #1)*

> Based on my educational background I never thought that scientific laws and theories are tentative and could change and consequently I ended up memorizing them. *(Participant #13)*

> It is only in the last year (when I started this graduate program) that I have come to understand that science is progressive and changing. Furthermore, scientists can think differently and interpret data in different ways. *(Participant #17)*

One of the best examples to illustrate the tentative nature of scientific knowledge, are the atomic models of Thomson, Rutherford, Bohr, and Bohr–Sommerfeld, which form part of science curricula and textbooks both at the high school and freshman university level, in almost all parts of the world (cf. Niaz & Cardellini, 2011). These models changed in quick succession over a short period of about 15 years (early twentieth century) and all of them were supposedly based on "objective" scientific knowledge (cf. Niaz, 2009a). Despite this it is difficult to convince teachers and students with respect to the changing and progressive nature of science. Philosopher of science, Ronald Giere (2006), has explained this in cogent terms, "no empirical claims can be known with absolute certainty to be true. It always remains possible we will find out later that we were earlier mis– taken" (p. 5). Indeed, with the postulation of the wave–mechanicalmodel of the atom (starting in the late 1920s), Thomson, Rutherford, Bohr, and Sommerfeld must have realized that they were "mistaken." This precisely shows the changing and progressive nature of science (cf. Lakatos, 1970).

6. Curricular Programs and Reforms (n = 8)

The following responses illustrate participants' thinking:

> In order to facilitate the inclusion of different aspects of the nature of sci– ence we shall have to reform the curriculum. This will help not only pre– and in–service teacher training but also facilitate a better appreciation of how science progresses among students and even perhaps the general pub– lic. *(Participant #10)*

> Our society is resistant to change and even the existing curricular structure of our university will not be receptive to the "new science" based on the different aspects of the nature of science. *(Participant #24)*

Reform documents in various parts of the world have not only called for adapting the curriculum to our current needs but explicitly espoused the inclu– sion of the HPS and consequently the need to revise the textbooks (AAAS, 1993; Project Beyond 2000 in the UK, Millar and Osborne, 1998). Jonathan Osborne (2007), co–authorof Project Beyond 2000, has expressed the need for change in succinct terms:

> Science education persists with presenting an idealized view of science as objective, detached and value free . . . [In contrast] study of the history of ideas and the evidence on which they are founded must lie at the core . . . [providing] opportunity to consider data which has no clear interpretation and to consider plural alternatives. *(pp. 175–179)*

7. Empirical and Not Theory-laden Nature of Science (n = 8)

The following responses illustrate participants' thinking:

> Inclusion of the different HPS–relatedaspects can weaken the supposedly "scientific convictions" of the teachers, as these depend only on experi– ments and empirical data. *(Participant #2)*

> Most teachers and students consider theories and laws to be synonymous, and this is not so even if we have additional empirical evidence. Furthermore, as all observations are theory–ladenit is possible that a scientist may have been subjective due to his presuppositions. *(Participant #3)*

These responses refer to the role of important aspects of the HPS, namely empirical evidence and that experimental data are theory–laden. The history of science shows that no amount of empirical evidence can unambiguously demonstrate the "truth" of a theory (Giere, 2006). Furthermore, as experimental data are difficult to interpret, we need plural alternatives (Osborne, 2007) and competitive cross–validationbased on peer review (Campbell, 1988b). These aspects if introduced appropriately in the classroom need not weaken the "scientific convictions" but, on the contrary, deepen our understanding of science.

8. Objectivity in Science Based on Experimenticism (n = 7)

This refers to aspects 4, 5, and 6 of the NOS (Reading 18, p. 411) and the following responses illustrate participants' thinking:

> It is difficult to teach our students about the nature of objectivity (aspects 4, 5, 6 in Reading 18) when there exists a deeply rooted tendency with respect to the scientific method being based on observation and experimentation. Furthermore, objectivity is "blind" and does not depend on the observer but rather is part of the nature of facts. *(Participant #1)*

> Many textbook authors present the "tentative" as "definitive," without explaining how the scientists arrived at a certain conclusion. This affects the introduction of new teaching strategies as knowledge is considered to be based on the opinion of a scientific authority and not on a process of competitive cross–validationthat requires critical peer evaluation. *(Participant #9)*

These aspects were perhaps the most novel and difficult to understand and still seven teachers referred to them. If science is "objective" as generally understood in the positivist perspective, then Newton's laws must have been the prime example of objectivity in science. However, at the beginning of the twentieth century, Einstein's theory of special and general relativity questioned Newton's perspective. Does this mean that Newton's laws were false and that he was not objective? (cf. Niaz, 2009a, pp. 14–16). Similarly, Tsai (2006) has argued for the inclusion of such aspects in teacher training: "The empiricist position assumes that scientific knowledge is a discovery of an *objective reality* external to ourselves and discovered by observing, experimenting or application of a scientific method" (p. 364, emphasis added). Let us now compare this with what Steven Weinberg (2001), Nobel laureate in physics, has to say about objective reality and truth in science: "What drives us onward in the work of science is precisely the sense that there are *truths out there to be discovered*, truths that once discovered will form a permanent part of human knowledge" (p. 126, emphasis added). Giere (2006) has characterized such philosophical positions as "objectivist realism" (p. 5) and explained cogently:

> Weinberg should not need reminding that, at the end of the nineteenth century, physicists were as justified as they could possibly be in thinking that classical mechanics was objectively true. That confidence was shattered by the eventual success of relativity theory and quantum mechanics a generation later. *(p. 118)*

9. Scientific Method (n = 6)

This refers to aspect 10 of the NOS (Reading 18, p. 411), and the following responses represent participants' thinking:

> All our educational literature is impregnated with the universal scientific method, which is more of a recipe and "Lord's Prayer" charged with empirico–positivistnotions that constitute "good science." *(Participant #1)*

> If a student in his dissertation states that, "there is no scientific method that provides a sequence of steps," then it may probably not be accepted as it is not in accord with university regulations for presenting a thesis. *(Participant #24)*

The scientific method constitutes an important part of the positivist framework of presenting good science. However, it is perhaps not so much the university regulations but, rather, the teachers themselves who are reluctant to abandon this legacy.

Note: In order to provide reliability of results (triangulation of data sources), participants' critical essays and interventions during the 18 class presentations were checked (see section Multiple data sources, for a pool of 510 written questions and comments). It was found that all participants expressed similar ideas on at least two different occasions.

Participants' Responses to Item 3

It is important to note that this item formed part of the Final exam, which was given in the last session of the final week of the course. In contrast, Items 1 and 2 formed part of the Initial exam, which was given during the first session of the final week. This means that between the two exams, participants had the opportunity to engage in 36 hours of classroom presentations, discussions, and question–answer sessions, which provided a considerable amount of experience and feedback with respect to the issues being discussed. This item provided a big challenge to the participants as they were required to suggest guidelines/strategies for solving the dilemma due to the tension between empirical data and "heuristic principles" (Schwab, 1962).

Based on participants' responses the categories shown in Table 5.3 were generated.

1. Motivation for Doing Research (n = 17)

The following responses represent participants' thinking:

> The solution lies in the use of articles from the research literature, in a way that the students can "discover" new ideas for themselves. For example, while teaching a particular topic we can provide appropriate materials in order to generate discussions, conflicts and opinions that can later be accepted by a majority of the students. *(Participant #10)*

> For a particular topic I can select a research article that provides a historical reconstruction of a particular experiment or episode. Next, I would present to the students (in small groups) the two versions: one from their traditional textbook and the other provided by the research article. This could lead to a discussion based on arguments and counter–argumentsand eventually facilitate conceptual change. *(Participant #15)*

> In a teaching strategy the new arguments must first attempt to reorient students' prior epistemological beliefs. In this context, it is important to understand the distinction between methodological aspects (empirical data) and the heuristic principles. It is important to make students aware that a majority of chemistry textbooks do not reflect this distinction. This

TABLE 5.3 Participants' Responses to Item 3 (How to solve the dilemma due to the tension between empirical data and "heuristic principle")

*Response category**	*Frequency (n)*
1. Motivation for doing research	17
2. The role of arguments and counter–arguments	15
3. The role of controversies among rival theories	9
4. Lakatosian perspective of conceptual change	7
5. The role of cognitive conflict	6
6. Understanding conceptual change beyond alternative conceptions	1

* These categories were generated based on participants' responses to Item 3.

> provides the background for students to critically appraise the textbooks. Just like Thomson and Rutherford, although professor and student but still had their differences, our students can also start to understand and even perhaps write the history of science. Furthermore, I have to make sure that the new arguments and counter–arguments are grounded in history and philosophy of science and easily accessible to the students. *(Participant #22)*

> First we have to convince ourselves that the "rhetoric of conclusions" is not the only or best strategy to facilitate students' conceptual change. One example of how change is possible is my own experience as I was trained under the guidance of textbooks that emphasize the "rhetoric of conclusions." Different courses in this graduate program have helped me to see the "other side of science," which is not presented in the majority of the textbooks. This has helped me to take decisions, have arguments in order to be the spokesperson of the "other side of science." This provides me an opportunity to convince my students with arguments (and their counter–arguments), that we have to go beyond the application of formulae and understand science within a social and historical context that facilitated the construction of these concepts. I have already started to introduce these ideas in my classroom while teaching the atomic models of Thomson, Rutherford and Bohr. In the case of Millikan's oil drop experiment we can even make some videos in order to understand evidence, like the television series "Crime scene investigation," which the students like very much. *(Participant #25)*

Interestingly, a majority of the participants suggested that based on the research literature the classroom itself could be the *scenario* of debates between the text–book presentations and the historical reconstructions of different episodes. If Rutherford, a student of Thomson, could disagree with his teacher, then given the appropriate research material our students could disagree/agree among them–selves and that can include the teacher as well. This can "liven," the classroom activities and students can feel free to accept one or the other interpretation and eventually reach a consensus.

2. The Role of Arguments and Counter-arguments (n = 15)

The following responses represent participants' thinking:

> Instead of a simple repetition of experimental details, interactive argumen–tation can enhance students' interest and motivation and thus facilitate con–ceptual change. *(Participant #5)*

> It is not a question of substituting one conception with another. On the contrary, we need to generate arguments that can extend, eradicate, and

> connect students' ideas with the new propositions. This will lead to controversies between the new and old ideas (rival hypotheses) and eventually facilitate conceptual change (Strike & Posner, 1992; Villani, 1992). *(Participant #6)*

> The fact that students often do not accept arguments and offer resistance to conceptual change is a consequence of how science is taught. Both teachers and textbooks in general follow a vision of science that is based on the scientific method that emphasizes: scientific knowledge is based on laws that are unequivocally based on a series of "invariably logical conclusions" that are, in turn, based on experimental results. This constitutes an obstacle (hence the need for counter-arguments)for any new teaching strategy. *(Participant #11)*

Again a majority of the participants emphasized the importance of generating arguments and counter-argumentsin the classroom (cf. Osborne, Erduran, & Simon, 2004). However, in order to facilitate conceptual change it is essential that the new arguments must in some sense "connect" with students' prior beliefs and thus the two could constitute rival hypotheses. This is a difficult task as students (and to compound matters, teachers as well), consider science to be a product of logical conclusions drawn from unambiguous experimental data that need not involve controversies or rival hypotheses.

3. *The Role of Controversies among Rival Theories (n = 9)*

The following responses represent participants' thinking:

> Students are generally not aware of the distinction between the empirical data and the heuristic principles. Hence it is essential to provide them information with respect to scientific practice based on controversies and competition between rival theories. *(Participant #2)*

> Classroom discussions must emphasize that when two or more theories clash based on basic principles and arguments, these can co-existfor some time and one is not necessarily replaced by the other. This will show to students that we do not have an absolute "truth" and that scientific theories change. *(Participant #24)*

These participants emphasized the role of scientific practice, and that science does not facilitate instant rationality (Lakatos, 1970). Rival theories can explain the same set of data, and the two can co-exist,while the scientific community deliberates based on peer evaluation.

4. Lakatosian Perspective of Conceptual Change (n = 7)

The following response represents participants' thinking:

> Obviously, if we want to introduce conceptual change based on heuristic principles, our present textbooks are not of much help. In practice, it appears that arguments based on historical reconstructions are not sufficient. This seems to be inherent in human nature and in our particular case due to an extensive tradition of presenting historical facts as "rhetoric of conclusions." However, historically all theories resist changes and seek protection from the "positive heuristic" (cf. Lakatos), until falsification. Perhaps, the same occurs in individual human beings, who try to protect the "hard-core" of their beliefs. If an individual does not resist conceptual change and easily accepts all the new ideas, then his cognitive structures would be in a chaos and it would be difficult to establish a firm base for problem solving. Consequently, conceptual change of necessity is a slow process and requires perseverance—a single drop falling on a stone will not break it, however a constant flow of drops over a prolonged period of time will. *(Participant #1)*

These participants explicitly referred to the Lakatosian framework and pursued the consequences of accepting new ideas with no resistance. In other words, acceptance of new ideas without any critical appraisal would not facilitate cognitive restructuring and hence any change at best would be superficial. Furthermore, an analogy is drawn between how scientists resist changes in their core beliefs by invoking the "positive heuristic" (also the protective belt) and students' rejection of new ideas.

5. The Role of Cognitive Conflict (n = 6)

The following response represents participants' thinking:

> Conceptual change theories are based among others on the ideas of Lakatos, Laudan and Toulmin, in which change is conceptualized as a process of evolution. If the new conception can help students to resolve more problems, then the ensuing cognitive conflict can facilitate conceptual change. *(Participant #17)*

The participants explicitly referred to the introduction of cognitive conflict within an HPS perspective (Posner, Strike, Hewson, & Gertzog, 1982; Strike & Posner, 1992). Similarly, Adey and Shayer (1994) have recognized the importance of cognitive conflicts within the Piagetian framework.

6. Understanding Conceptual Change beyond Alternative Conceptions (n = 1)

Only one participant referred to this important issue in the following terms:

> It is important to note that in some instances, it is not a question of change, but rather a complete absence of any prior beliefs with respect to a particular topic (economy of cognitive thought). This provides the student an opportunity for significant learning by not following traditional learning "pitfalls" but rather going beyond and "constructing" new ideas. *(Participant #16)*

This suggests that not all new ideas may face resistance or the same degree of resistance. This situation can easily be used to advantage by the teacher provided adequate teaching strategies are designed.

Note: In order to provide reliability of results (triangulation of data sources), participants' critical essays and interventions during the 18 class presentations were checked (see section Multiple data sources, for a pool of 510 written questions and comments). It was found that all participants expressed similar ideas on at least two different occasions.

Conclusions and Educational Implications

Item 1 in this study asked participants to select five aspects of this course that they found novel. Participants considered the following aspects to be novel: (a) The role of the HPS in science education (n = 24); (b) The importance of the NOS (n = 24); (c) Conceptual change in learning science (n = 18); (d) Students' alternative conceptions (n = 13); (e) Critical and reflective thinking (n = 13); (f) Research articles for questioning inadequate ideas of teaching science (n = 13); (g) Motivation for doing further research (n = 10); (h) Heuristic principles and rhetoric of conclusions (n = 9); (i) Science is objective only in a certain context of scientific development (n = 9); and (j) Cognitive psychology for meaningful learning (n = 5). Most of the participants considered the inclusion of the HPS and NOS as novel aspects, in contrast to their previous experience based on the empiricist perspective of science. Some of the participants were surprised as to how textbooks and the curricula have simply ignored the historical background of experiments, which is so fundamental to understanding science. The NOS was considered to be an integral part of science, which involves among other aspects, various controversies, such as: Thomson–Rutherford, Mendeleev and the placing of argon, and Millikan–Ehrenhaft. Participants recognized the need for going beyond algorithmic problem solving and introducing conceptual change teaching strategies. Some participants considered students' alternative conceptions as part of the hard-core of beliefs and, consequently, that conceptual change is difficult and requires strategies that question the hard–core (Lakatos, 1970). Participants invoked the Lakatosian perspective by considering students' alternative conceptions and its criticism (based on the scientific conception) as rival or competing models. Adaptation of research articles (e.g., related to alternative conceptions, textbooks, etc.) for use in the classroom was considered to be a novel aspect of this course. Having recognized the importance of research articles, participants manifested motivation for doing further research themselves and thus innovate teaching of science. For most chemistry textbooks and curricula it is the handling of experimental apparatus that facilitates success and makes science objective. Most participants understood that this perspective represents the "experimenticist fallacy" (Holton, 1969), that leads the textbooks to present a "rhetoric of conclusions" and supposedly makes science "objective." On the contrary, objectivity in science is the consequence of a social process of "competitive cross–validation"that requires interpretation of experimental data based on rival sets of "heuristic principles."

Having reflected on the novel aspects of this course, Item 2 asked participants to consider factors (within the context of the NOS) that can impede the implementation of new teaching strategies. Participants considered the following factors to be possible obstacles: (a) Empiricist presentations in chemistry textbooks that lack an HPS perspective (n = 19); (b) Schwab's "rhetoric of conclusions" and lack of "heuristic principles" (n = 18); (c) Teachers' and students' epistemological views (n = 16); (d) Scientific progress devoid of controversy, interpretation of data, and idealization (n = 14); (e) Non–recognitionof the tentative nature of scientific knowledge (n = 11); (f) Curricular programs and reforms (n = 8); (g) Empirical and not theory–ladennature of science (n = 8); (h) Objectivity in science based on experimenticism (n = 7); and (i) Scientific method (n = 6). Most of the participants considered the empiricist framework of textbooks as a major obstacle in the implementation of new teaching strategies. Instead of presenting science as a finished product (final form) it would be more motivating for both students and teachers if science is presented as a human enterprise in which the difficulties involved in constructing theories are highlighted (Niaz, 2009a). Participants considered their own and students' epistemological views as another important obstacle. In general, teachers feel comfortable in teaching as they have been taught, and hence any change will require considerable time and effort. It is difficult to present a critical perspective in the classroom if teachers do not have access to how scientists go about doing science (Niaz, 2010a). The tentative nature of scientific knowledge is an important aspect of the NOS and it can be easily illustrated within topics (e.g., atomic models) that already form part of the science curriculum. Understanding science requires the differentiation between the empirical data and the scientists' presuppositions (theory–laden),which inevitably leads to alternative interpretations. Some participants referred to the conceptualization of objectivity as part of the "nature of facts," that is, "truths are out there to be discovered." For some participants the scientific method continues to be a good indicator of doing "good science."

After having considered the novel aspects of the course (Item 1), and factors that can impede new ideas (Item 2), as a next step Item 3 provided participants the opportunity to suggest new teaching strategies, such as: (a) Motivation for doing research (n = 17); (b) The role of arguments and counter–arguments(n = 15); (c) The role of controversies among rival theories (n = 9); (d) Lakatosian perspective of conceptual change (n = 7); and (e) The role of cognitive conflict (n = 6). A majority of the participants recognized the need for creating a new *scenario* in the classroom in which students and teachers could debate the different historical reconstructions based on research articles. These debates based on new ideas must correlate with students' prior epistemological beliefs, so that the two could constitute rival hypotheses. Furthermore, rival theories can explain the same set of data differently, and the two can co–exist,and it is the "competitive cross–validation" through peer review that facilitates understanding. Drawing on the Lakatosian perspective it is plausible to suggest that resistance to conceptual change is understandable and of necessity any change requires prolonged interaction, which may lead to cognitive conflicts that eventually provide a better explanation.

Finally, this study has the following educational implications:

1. Inclusion of the NOS based on an HPS perspective in contrast to the empiricist perspective of most textbooks and classroom environments.
2. The importance of "heuristic principles" in order to understand different aspects of the NOS, as part of the historical episodes.
3. Using the Lakatosian perspective that by considering students' alternative conceptions and its criticism (based on the new ideas) as rival or competing models can facilitate conceptual change. History of science shows that scientists themselves generally experience a similar process in which rival theories can co–existand explain the same set of data differently, until a consensus is achieved through peer interaction.
4. In order to make science "objective," textbooks follow the "experimenticist fallacy" which leads science content to be presented as "rhetoric of conclusions." On the contrary, objectivity in science is a consequence of a social process that requires interpretation of experimental data based on rival sets of "heuristic principles."
5. Most textbooks conceptualize "objectivity" as part of the "nature of facts," that is, "truths are out there to be discovered." It would be more motivating for students and teachers, if science was presented as a human enterprise in which the difficulties involved in constructing theories are highlighted.
6. There is a need to create a new *scenario* in the classroom in which students and teachers could present arguments and counter–argumentsbased on historical reconstructions of various episodes in the history of science.
7. It is interesting to compare the two lists of NOS aspects presented in this chapter. The first list appears in the Introduction section and has 11 aspects. The second list appears in Item 2 (based on Niaz, 2005b), and has 10 aspects.

In most respects the two lists deal with the same NOS aspects. However, the second list has some novel features based on objectivity (aspects 4, 5, and 6) and how it can be conceptualized in scientific progress. Furthermore, the NOS aspects in both lists have been "anchored" in actual episodes from the history of science, and thus can also be considered as domain–specific.

8. A reconstruction of various historical episodes reveals that the positivist perspective in which objectivity leads to finding "truths out there to be discovered" is questionable and misleading.

6

HOW THE VIEWS OF LEON COOPER (NOBEL LAUREATE) CAN INFLUENCE IN-SERVICE TEACHERS' UNDERSTANDING OF THE NATURE OF SCIENCE

Introduction

Recent research has recognized the importance of the nature of science (NOS) for science education (Clough & Olson, 2008; Hodson, 2009). An adequate understanding of the NOS facilitates students' and teachers' comprehension of 'science in the making,' based on historical, philosophical, and epistemological considerations. A considerable amount of research has been conducted to explore students' and teachers' understanding of NOS views in different parts of the world (Blanco & Niaz, 1997; Dagher & BouJaoude, 2005; Dogan & Abd-El-Khalick, 2008; Kang, Scharmann, & Noh, 2005; Liu & Lederman, 2007; Liu & Tsai, 2008; Sadler, Chambers, & Zeidler, 2004; Southerland, Johnston, & Sowell, 2006). Science education research has also devoted considerable efforts towards teaching the NOS in the classroom (Abd-El-Khalick& Akerson, 2004; Ford & Wargo, 2007; Khishfe & Lederman, 2007; Niaz, 2011; Wong et al., 2008). Similarly, there seems to be consensus that teaching the NOS needs a reflective, explicit, and activity-basedapproach (Smith & Scharmann, 2008). For further details, see Chapter 3.

Rationale of the Study

Gooday, Lynch, Wilson, and Barsky (2008) have asked a very interesting question: "Have any Nobel Prize winners declared that history of science was the secret of their creative success?" (p. 322). These authors (Kenneth G. Wilson, among them, himself a Nobel laureate in physics) then responded in the following terms: "Well—not yet, anyway" (p. 322). Without waiting for a Nobel laureate to provide information with respect to their query, Gooday et al. (2008) have stated cogently:

> The key role of history here is characterizing the complexities of how science *changes*. So many science textbooks unhelpfully—and above all inaccurately—cultivate a rather static image of scientific disciplines, as if they were completed with comprehensive certainty. It is perhaps not difficult to understand how this gross simplification might arise as the result of a pedagogical need to "tidy up" the presentation of science to meet the needs and capacities of students. But faced with the textbook spectacle of such an apparently unalterable monolith, is it any wonder that students can have difficulty conceiving how they might ever contribute to science? *(p. 326, original italics)*

This raises important issues of how our present day science textbooks depict science as an "unalterable monolith" and, consequently, students may not feel motivated to contribute towards changing science. Research in science education has shown that inclusion of an HPS framework in our textbooks can provide students with a glimpse of how scientific knowledge is tentative and thus requires the contribution of future generations of students (Abd–El–Khalick,Waters, & Le, 2008; Niaz, 2008b; Stinner, 1992). Another Nobel laureate has cautioned against students memorizing laws and instead recommended that, "we convey to the students the dialectic by which scientific discoveries are made and recognized" (Gell–Mann,1971, p. 25).

Leon N. Cooper, recipient of the 1972 Physics Nobel Prize for his work on superconductivity, has included the history of science in his general physics textbooks (e.g., Cooper, 1970). Rodríguez and Niaz (2004b) have analyzed the presentation of atomic structure in general physics textbooks published in the USA, and found that Cooper's (1970) textbook was the only one that satisfactorily presented the atomic models of J.J. Thomson, E. Rutherford, and N. Bohr within an HPS perspective. One of the most important features of Cooper's (1970) presentation was the extensive use of original historical sources. Of the 41 textbooks analyzed by Rodríguez and Niaz (2004b), Cooper (1970) had the highest score of 13 points (out of a maximum of 16), and the second highest score was 4 points. Given Cooper's interest in presenting general physics within an HPS perspective, Niaz, Klassen, McMillan, and Metz (2010b) invited him to participate in an interview study based on 13 questions, to which Cooper responded in writing. In general, this study found that Cooper endorsed the historical perspective in teaching science and that most textbook authors have a fundamental premise, that is, the empiricist epistemology in which progress in science is presented as a simple accumulation of experimental data, rather than an attempt to understand and interpret data. Based on Niaz et al. (2010b), this study elaborated a five–item questionnaire, in order to explore in–serviceteachers' understanding of the NOS. The objective of this study is to evaluate how the views of a Nobel laureate can influence in–serviceteachers' understanding of the NOS based on a reflective and explicit, activity–basedapproach.

Method

This study is based on the responses of 20 in–serviceteachers who had enrolled for the following required course: "Science, Technology, Ethics and Creativity in Research," as part of their Doctoral degree program in Materials science, at a major university in Latin America. The course was sub–dividedinto the following six topics: (1) Introduction; (2) Scientific development: A history and philosophy of science perspective; (3) Methods and creativity in research; (4) Structure of a scientific publication; (5) Review of research and the use of internet; and (6) Evaluation of the efficiency of research based on good lab practice. The study reported here was based on Topic 2 and the author of this study was the instructor. Participants' undergraduate degree ("Licenciatura" or equivalent) was in the following fields: Chemistry = 7, Physics = 6, Petroleum engineer = 2, Chemical engineer = 1, Mechanical engineer = 1, Metallurgical engineer = 1, Systems engineer = 1, and Materials science = 1. The age of the participants ranged from 25 to 40 years and the teaching experience ranged from 5 to 15 years (female = 12, male = 8). All participants were working in universities and colleges and most of them were giving freshman courses. Some of them already had a Master's degree and publications in international research journals in their respective fields. Few of the participants had basic knowledge of the work of Popper, Kuhn, Lakatos, Giere, and other philosophers of science.

Course Outline

All participants were provided with the following course outline, one month before the participants met with the instructor:

> This course topic is designed to facilitate an understanding of scientific development based on a history and philosophy of science perspective (Popper, Kuhn, Lakatos, Feyerabend, Laudan, Cartwright, Giere, and others). Among others, the following experiments will be discussed:
>
> 1. Michelson–Morley experiment to determine the velocity of the earth with respect to the ether (1887).
> 2. J.J. Thomson's cathode ray experiments (1897).
> 3. E. Rutherford's alpha particle experiments (1911).
> 4. N. Bohr's understanding of experiments that determined spectra of elements (1913).
> 5. R. Millikan's (1910–25) oil drop experiment to determine the elementary electrical charge (electron).
> 6. Experiments related to the photoelectric effect as evidence for Einstein's hypothesis of lightquanta (1916).
> 7. The bending of light in the 1919 eclipse experiments as evidence for Einstein's theory of general relativity.

8. Martin Perl's (1997) experiments to isolate "quarks" at the Stanford Linear Accelerator Center, SLAC (Perl was awarded the physics Nobel Prize in 1995).

A history and philosophy of science perspective reveals that in these experiments (and others), scientific development is characterized by: (a) Experiments are important, however, their interpretations are even more important; (b) Rivalries and conflicts between scientists have repercussions in the development of their theories; (c) At times, the same data can be explained/interpreted through different models and theories; (d) Scientific theories are tentative; (e) Scientists do not inevitably abandon their theories on finding anomalous data; and (f) The scientific method does not necessarily play an important part in scientific research.

Reference:

Niaz, M. (2009a). *Critical appraisal of physical science as a human enterprise: Dynamics of scientific progress*. Dordrecht, the Netherlands: Springer.

Reading material:

a. Niaz, M. (2010a). Are we teaching science as practiced by scientists? *American Journal of Physics, 78*(1), 5–6.
b. Niaz, M., Klassen, S., McMillan, B., & Metz, D. (2010b). Leon Cooper's perspective on teaching science: An interview study. *Science & Education, 19*(1), 39–54. (Cooper was awarded the physics Nobel Prize in 1972, for his contribution to the theory of superconductivity).

Note: All participants were provided with copies of the two articles in the reading material and they were compulsory reading. The reference was not required reading, however, a copy of the book was available in the university library.

Course Organization and Activities

Participants had the opportunity to study the reading materials during the first four weeks of the course and also communicated among themselves and with the instructor through emails. At the end of the fifth week, participants met with the instructor for a four–hoursession, which included the following activities: (a) A formal PowerPoint presentation by the instructor, based on eight experiments referred to in the Course outline (60 minutes); (b) Interactive discussion based on the presentation and other aspects of the course (30 minutes); (c) Recess (15 minutes); (d) Interactive discussion with respect to various aspects of the course material (30 minutes); and (e) Evaluation using a 5–itemquestionnaire based on Cooper's views on teaching science (90 minutes). This evaluation was open–book and participants could use any materials or personal notes that could be of help. The questionnaire is presented in the next section.

Evaluation

All participants responded to the following 5–itemquestionnaire, which formed part of their formal evaluation for this course:

> Consider the following assertions based on Leon Cooper's perspective on teaching science (as reported in Niaz et al., 2010b). In each case you are asked to respond by indicating and explaining, whether you are: (a) In agreement; (b) In partial agreement; or (c) In disagreement:
>
> 1. Varying interpretations of observations do not undermine the objective nature of science.
> 2. New ideas in science comprise an imposed vision of the world, and these ideas are then slowly accepted by the scientific community.
> 3. The current view in any science is almost always a mixture of data, hypotheses, theoretical ideas, and conjectures.
> 4. Since experiments are difficult to perform and understand, scientists rely on their presuppositions to guide the integration of data, theory, and conjectures.
> 5. Case studies based on various experiments show that scientists use intuition which is guided by facts, conjectures, and speculations.

All five assertions are based on a specific context and questions as reported by Niaz et al. (2010b):

- Assertion #1 is based on Question 7, which deals with the varying interpretations of observations, in the context of quantum theory, general relativity, and Newtonian theory.
- Assertion #2 is based on Question 8, which deals with "vision of the scientist," in the context of Newtonian mechanical notion of the universe, social Darwinism, quantum theory, and impressionism (style of painting).
- Assertion #3 is based on Question 10, which deals with the relationship between experiments and conjectures, in the context of E. Rutherford's alpha particle experiments.
- Assertion #4 is based on Question 9, which deals with the relationship between an experiment and the scientist's presuppositions, in the context of J.J. Thomson's model of the atom and string theories.
- Assertion #5 is based on Question 13, which deals with the role of intuition guided by facts, conjectures, and speculations in cutting–edgeexperimental

work, in the context of Martin Perl's philosophy of speculative experiments for isolating quarks.

At this stage it is important to note the following aspects of the course organization and activities: (a) Both the reading material and class room discussions provided participants with an overview of 'science in the making'; (b) Due to their physical science background, all participants were aware of the context on which the assertions were based; (c) Participants were exposed to an HPS–based discussion of the context and the assertions in the reading materials; (d) The importance of understanding experiments within an HPS perspective was discussed explicitly in the presentation by the instructor at the end of the fifth week (before the participants responded to the questionnaire); and (e) It is plausible to suggest that this study is based on a "reflective, explicit, activity–basedapproach" (cf. Akerson, Abd–El–Khalick& Lederman, 2000; Khishfe & Lederman, 2007; Smith & Scharmann, 2008).

Results and Discussion

On all five assertions, all responses were evaluated according to the following criteria: (a) An Agree response was considered to be an "Informed" view of the NOS; (b) Partially agree was considered to be a "Partially Informed" view of the NOS; and (c) Disagree was considered to be a "Traditional/Naive" view of the NOS. It can be observed that a majority of the participants agreed (ranging from 60 to 80%) with Assertions #1, 3, 4, and 5 (See Table 6.1), considered to be an "Informed" view of the NOS. Only five (25%) participants agreed with Assertion #2, which had also the highest number (n = 12) of the "Partially Informed" view of the NOS. On all five assertions, very few participants had a "Traditional/Naive"' view of the NOS. Next, I present a discussion and examples from participants' responses to each of the five assertions.

Research Assertion #1 (Varying Interpretations of Observations do not Undermine the Objective Nature of Science)

This research assertion is based on Question 7 (Niaz et al., 2010b, pp. 46–47) in which Cooper explains how our interpretations of phenomena change over time in the context of quantum theory, general relativity, and Newtonian theory. Following are four examples of explanations provided by participants who "Agreed" (Response (a)) with this assertion:

> This assertion represents the richness of the contributions of science. The objective of science is to look for models that approximate to reality—a reality not based on just one vision of what is observed, but on the contrary a diversity of visions.*(Participant #1)*

TABLE 6.1 Participants' understanding of the nature of science based on Cooper's[1] perspective of teaching science (n = 20)

Item	*Number of participants' responses*		
Research assertion	*a*[2]	*b*[3]	*c*[4]
1. Varying interpretations of observations do not undermine the objective nature of science	14 (70%)	4 (20%)	2 (10%)
2. New ideas in science comprise an imposed vision of the world, and these ideas are then slowly accepted by the scientific community	5 (25%)	12 (60%)	3 (15%)
3. The current view in any science is almost always a mixture of data, hypotheses, theoretical ideas, and conjectures	15 (75%)	5 (25%)	0 (—)
4. Since experiments are difficult to perform and understand, scientists rely on their presuppositions to guide the integration of data, theory, and conjectures	12 (60%)	7 (35%)	1 (5%)
5. Case studies based on various experiments show that scientists use intuition which is guided by facts, conjectures, and speculations	16 (80%)	3 (15%)	1 (5%)

Notes:

1 Cooper's perspective on teaching science is based on Niaz et al. (2010b)

2 Response (a) (Agree) was considered as an "Informed" view of NOS

3 Response (b) (Partially agree) was considered as a "Partially Informed" view of NOS

4 Response (c) (Disagree) was considered as a "Traditional/Naive" view of NOS

> We have observed that this happens very frequently in science, and forms a part of the nature of science. For example, the controversy between Thomson and Rutherford—same experimental results and different interpretations. *Participant #9)*

> This assertion corresponds with reality. Scientists interpret observations as a function of their previous knowledge and what is of interest to them. In the development of science this happens frequently—different explanations are presented for the same phenomenon, leading to conflicts, which eventually facilitate consensus and an ample vision of reality. *(Participant #14)*

> Depending on her/his beliefs, background and presuppositions a scientist can provide different interpretations for the same observations—this precisely constitutes an opportunity to do science. *(Participant #18)*

The following is an example of an explanation provided by a participant who partially agreed (Response (b)) with this assertion:

> Science is done by persons, each one of them has a criterion and consequently a different point of view. Thus science must have objective criteria in order to know which one of these represents the truth. *(Participant #6)*

The following response is an example of an explanation provided by a participant who disagreed (Response (c)) with this assertion:

> The objectivity of science cannot be affected by different interpretations of the same observations, as the logic of science would impose itself on the incorrect interpretations. *(Participant #5)*

It is interesting to observe that participants who agreed (Response (a)) with this assertion, emphasized that different interpretations of observations provide an "ample vision of reality" as a means to elaborate/construct models that "approximate to reality." In contrast, the participant who partially agreed (Response (b)), referred to "objective criteria" that could help us to find "the truth." The participant who disagreed (Response (c)) explicitly referred to the "logic of science" that would help to discard incorrect interpretations. There seems to be a fundamental difference between these responses: Participants who agreed with the assertion seem to be suggesting that the "ample vision" provided by the different interpretations facilitates an understanding that helps us to approximate reality. In contrast, participants who partially agreed or disagreed with the assertion seem to be suggesting that it is not the "ample vision" (based on conflicts, interactions among scientists) but, rather, the "objective criteria" and "logic of science" that help us to find the truth (correct interpretation).

Research Assertion #2 (New Ideas in Science Comprise an Imposed Vision of the World, and these Ideas are then Slowly Accepted by the Scientific Community)

This research assertion is based on Question 8 (Niaz et al., 2010b, pp. 47–48) in which Cooper goes to considerable length to explain his idea of the "vision of the scientist" framed in the context of Newtonian theory, quantum theory, social Darwinism, and impressionism. The analogy between a style of painting (impressionism) and progress in science is thought provoking, as in the beginning these painters were "accused of not being able to see things as they are." Another example that illustrates this aspect of the nature of science is the harsh criticism and rejection for many years by Millikan (1916) of Einstein's explanation of the photoelectric effect. Einstein provided a new vision (based on lightquanta) to understand the photoelectric phenomenon, whereas Millikan was a strong adherent of the classical theory of light (see Chapter 2). In other words, Millikan, despite being a brilliant scientist, resisted any change in his theoretical framework based on an "imposed" new vision of the world (for details see Niaz, 2009a, Chapter 8). It is interesting to note that only five participants had an Informed view (Response (a)) on this assertion, whereas 12 had a Partially Informed view (Response (b)).

The following is an example of an explanation from a participant who agreed (Response (a)) with this assertion:

> Situating in the historical context of science we find that scientists have always developed new concepts, some of which were controversial in their epoch and caused disagreements and rejections. However, with the passage of time other scientists come to accept theories that were previously rejected. *(Participant #14)*

The following are three examples of explanations from participants who partially agreed (Response (b)) with this assertion:

> In my opinion, a new idea or contribution does not necessarily have to be an imposed vision of the world. In the first place it needs to be evaluated by the scientific community, as this provides validity. However, this can vary depending on the context, time, and situation. *(Participant #15)*

> The essence of science is impregnated with liberty. If a vision of the world is imposed it does not provide opportunities for new ideas to emerge. However, I agree that a new idea is not easily accepted by the scientific community. *(Participant #18)*

> I agree that some new ideas are resisted in the beginning as they contradict our world of knowledge, and with the passage of time these are eventually accepted. However, I do not believe that these are an imposed vision of the world—this is how I understand it. *(Participant #19)*

The following response is an example of an explanation from a participant who disagreed (Response (c)) with this assertion:

> Science is not an imposed vision of the world. Science is knowledge and as such requires a structure: observation, experimentation, enunciation and verification of laws, for it to be accepted. *(Participant #17)*

It is interesting to observe that a majority of the participants partially agreed (Response (b)) with this assertion, and provided reasons as to why new ideas in science are not an imposed vision of the world, such as: new ideas first need to be evaluated by the scientific community; the libertarian ethos of science does not accept imposed views; and new ideas are always resisted. The history of science shows that new ideas are not only resisted but actively scrutinized by the scientific community and often lead to bitter controversies based on apparently plausible arguments. Of course, new ideas are evaluated by the scientific community. This evaluation, however, is based on competitive cross–validationof ideas that involves peer pressure and considerable rivalry. Again, it is not that scientists have any special means for imposing their ideas. For example, in the Millikan–Ehrenhaft controversy (determination of the elementary electrical charge, cf. Niaz, 2009a, Chapter 7), the controversy lasted for over 10 years, the scientific community itself was divided, and despite that Millikan was awarded the Nobel prize in 1923. Release of the Nobel Prize Archives shows that:

> [A]lthough Millikan was nominated for the prize in physics regularly from 1916 on, Svante Arrhenius, in the report he prepared on Millikan's work for the deliberations of the committee, noted as late as 1920 that even though most physicists had come to agree with Millikan in the dispute with Ehrenhaft, the matter was not yet resolved, and that Millikan should therefore not then be recommended for the prize. *(Holton, 1988, p. 196)*

Ehrenhaft, for his part, was adamant that Millikan's data did not conclusively demonstrate the existence of the elementary electrical charge, and continued to publish on the subject for many years after 1923. It is in such situations that Millikan's interpretation constituted an "imposed" vision of the scientist and it took a few more years (after being awarded the Nobel Prize) for the scientific community to definitely accept his thesis. Cooper's reference to the impressionists provides a good perspective to understand this facet of the NOS. Course material and class discussions provided various examples of such controversial episodes (e.g., the Thomson–Rutherford controversy lasted for many years and Thomson wielded considerable influence in the scientific community; despite this, in the long run, the community favored Rutherford's thesis). Modern philosophers of science have recognized the importance of controversies in scientific progress and, at the same time, pointed out their paradoxical role:

> Many major steps in science, probably all dramatic changes, and most of the fundamental achievements of what we now take as the advancement or progress of scientific knowledge have been controversial and have involved some dispute or another. Scientific controversies are found throughout the history of science. This is so well known that it is trivial. What is not so obvious and deserves attention is a sort of *paradoxical dissociation* between science as actually practiced and science as perceived or depicted by both scientists and philosophers. While nobody would deny that science in the making has been replete with controversies, the same people often depict its essence or end product as free from disputes, as the uncontroversial rational human endeavor par excellence. *(Machamer, Pera, & Baltas, 2000, p. 3, italics added)*

No wonder, if scientists and philosophers themselves are faced with a "paradoxical dissociation," both students and in–serviceteachers have considerable difficulty in understanding such issues. In this context it is interesting to note that the explanation provided by one of the participants (#17, reproduced above) who disagreed (Response (c)) approximates the traditional scientific method.

Research Assertion #3 (The Current View in any Science Is almost always a Mixture of Data, Hypotheses, Theoretical Ideas, and Conjectures)

The following are five examples of explanations from participants who agreed (Response (a)) with this assertion:

> Science is like a tree that grows by taking nutrients from data, hypotheses, ideas and conjectures. This consolidates its base and thus sustains with the passage of time. *(Participant #6)*

> Definitely, important scientific findings are based not only on experimental data, but also on beliefs, presuppositions, and conjectures. *(Participant #7)*

> Science is not based in being just in the laboratory and doing experiments. It is good to go to the lab with certain ideas, conjectures, etc. *(Participant #9)*

> In contrast, it is generally believed that science is based only in the verification of a hypothesis through experiments that facilitate data, which are then compared with what the scientist expected in order to postulate theories. *(Participant #14)*

> From Cooper's perspective these are fundamental elements of the epistemic matrix. *(Participant #18)*

The following are four examples from explanations provided by participants who partially agreed (Response (b)) with this assertion:

> This contrasts with what we are shown during our educational career, as we are practically induced to be dogmatic. On the other hand, I consider that we need to be critical and philosophers at the same time. *(Participant #1)*

> Although the image of the scientific method still lingers. However, new visions of what science is have started to emerge. *(Participant #8)*

> Besides, science must break with the empirical and project itself towards new tendencies, including its history. *(Participant #10)*

> Our present day textbooks project science as a rigorous and systematic demonstration of how the events occurred, and not as a result of the conjectures and presuppositions of the scientists who devoted their lives to science. *(Participant #12)*

At this stage it is important to recall that Research Assertion #3 is based on Question 10, in which Cooper referred to experiments and conjectures in the context of Rutherford's alpha particle experiments and stated explicitly: "This view of the world is based on actual data, on people's beliefs as to what the data means, and on conjectures about data that is not presently available" (reproduced in Niaz et al., 2010b, p. 49). All participants were familiar with Rutherford's alpha particle experiments. The novel feature of this assertion was that Cooper juxtaposed *experiments* with *conjectures*, and most participants found this quite intriguing if not challenging. Fifteen (75%) participants argued cogently (Response (a)) with respect to how science is not just doing experiments in the lab but also requires the inclusion of an epistemic matrix (Participant #18), based on Cooper's perspective. Five (25%) participants partially agreed (Response (b)) with the assertion, and provided very interesting explanations. Participant #1, posed a dilemma in being dogmatic and critical/philosophical; Participant #8, alluded to a contradiction between the scientific method and the new vision of science; Participant #10, pointed out the need for breaking with the empirical, based on a historical perspective; Participant #12, contrasted science as presented in textbooks (systematic and rigorous) and that presented by the scientists (conjectures and presuppositions). Although these four participants selected Response (b), it is plausible to suggest that these could easily have been classified as having an "Informed view" of the NOS (Response (a)).

Research Assertion #4 (Since Experiments are Difficult to Perform and Understand, Scientists Rely on their Presuppositions to Guide the Integration of Data, Theory and Conjectures)

This assertion is based on Question 9 (Niaz et al., 2010b, pp. 48–49) and dealt with the design of an experiment and the underlying theoretical framework, in the context of J.J. Thomson's cathode ray experiments. All participants were familiar with Thomson's experiments and his model of the atom. In his response, Cooper stated explicitly: "One of the values of a theoretical structure is that it focuses on the questions that are relevant" (reproduced in Niaz et al., 2010b, p. 48). Cooper then goes on to point out the importance of doing experiments to understand string theories. Interestingly, research in string theory has assumed a prominent place in theoretical physics, and still there is little direct empirical evidence for this idea (cf. Dawid, 2006).

The following are four examples of explanations from participants who agreed (Response (a)) with this assertion:

> An experiment cannot be performed without having an idea with respect to the expected result. This is one form of doing science: study what needs to be found and then verify through experiment, and thus integrate the two. *(Participant #4)*

> Scientists have almost blind faith that experimental data would guide them towards what they want to know. Consequently, based on their presuppositions they obviate data that do not coincide with their framework. *(Participant #7)*

> Surely, every scientist trusts her/his presuppositions. However, scientists do not possess absolute truth and hence the necessity for winning controversies based on peer reviewed publications. *(Participant #10)*

> To be sure, it would be impossible to design an experiment if we do not know what we want to know. It is precisely in this context that presuppositions play an important role. *(Participant #12)*

The following are three examples of explanations provided by participants who partially agreed (Response (b)) with this assertion:

> Certainly, when you go to the lab one always has a presupposition with respect to what can be obtained and frequently this deemphasizes other results that could have been interesting for further development and innovation. *(Participant #14)*

> Actually, it depends on the experiment and the degree of comprehension the scientist has about the subject. Furthermore, not all experiments are difficult. *(Participant #15)*

> Although, it should not be like this, as science must be exact in any lab and reproduce the results. Scientists do repeat experiments until they obtain the desired results. However, I do not agree with falsifying data—in such cases it is preferable not to publish. *(Participant #19)*

The following response is an example of an explanation provided by a participant who disagreed (Response (c)) with this assertion:

> Presuppositions of results can lead a researcher to commit some type of fraud, based on manipulation or invention of data. *(Participant #3)*

A majority of the participants agreed ($n = 12$) with this assertion and emphasized the following: Despite presuppositions, experimental data is still important (Participant #7), and the difficulties involved in designing experiments without knowing what we want to know (Participants #4 and #12). Participants who partially agreed (n = 7) expressed concern for the reliance on presuppositions (theoretical framework) and pointed out that: Science must be exact, results must be reproducible, data should not be falsified, and not all experiments are difficult.

These are important issues and have been debated extensively in both the philosophy of science and science education literature. In this context, it would be interesting to consider the controversy with respect to Millikan's oil drop experiment (Holton, 1978; Niaz, 2005a). This experiment is difficult to perform even today and it is not easy to reproduce Millikan's original results (Heering & Klassen, 2010; Jones, 1995; Klassen, 2009; Kruglak, 1972). For example, Kapusta (1975) found that some of the drops have much higher velocities than others, and so should one choose to observe the fast drops, the slow drops, or those of intermediate velocities? Most teachers ignore the fact that this was precisely Millikan's dilemma and a perspective based on the HPS can help to resolve it. In a recent study, Klassen (2009) used an HPS perspective to perform the experiment with undergraduate physics students. Interestingly, after the experiment, one of the students reported: "By having a preconceived notion of what *e* [charge of the electron] should be, we knew what to expect and disregarded observations that were not expected. In saying so, I believe Millikan depended on his preconceived notion as much as we did" (reproduced in Klassen, 2009, p. 604).

In the present course, when the oil drop experiment was discussed in the context of the Millikan–Ehrenhaft controversy, one of the students related her/his personal experience: I performed Millikan's oil drop experiment as an undergraduate student and have been a lab instructor for many years in which this experiment was included. I must confess that for so many years I did not understand how Millikan got his results, and it is only in the context of this course that I have understood the underlying dilemma (paraphrased from author's class notes).

It is important to note that the HPS perspective presented in this chapter does not endorse a "high view" of theory–ladenness,according to which experimental data are not important. On the contrary, setting up an experiment is difficult and expensive and interpretation of the data even more difficult. It is precisely due to these reasons that scientists generally seek guidance from their presuppositions and peer interaction. Various episodes from the history of science bear witness to this tension between experimental data and presuppositions (cf. Chapter 2 for details). Chapter 5 deals with the same problem in the context of classroom practice and suggests possible alternatives.

Research Assertion #5 (Case Studies Based on Various Experiments Show that Scientists Use Intuition which is Guided by Facts, Conjectures, and Speculations)

This research assertion is based on Question 13 (Niaz et al., 2010b, pp. 51–52) that referred to Martin Perl's (Nobel laureate in physics, 1995) philosophy of speculative experiments (search for quarks), in the context of Cooper's apprecia–

tion of Perl. In other words, in this case the participants were evaluating the views of two Nobel laureates in physics. Following are seven examples of explanations provided by participants who agreed (Response (a)) with the research assertion:

> In the case of cutting-edgeresearch, based on previous findings a scientist can have an intuition with respect to future results. Consequently, the scientist can help to make the field move forward by speculating and making conjectures. *(Participant #4)*

> It has been demonstrated that Millikan, 1910–1920 (and other scientists), adjusted his results to obtain the value of the charge of the electron, based on the value suggested by E. Rutherford. *(Participant #5)*

> A scientist must have imagination to see beyond what others cannot see. Consequently, he/she can use his intuition guided by reason and facts. *(Participant #6)*

> Various experiments in this course have demonstrated that scientists are human, and as such are subjective and orient their research based on their beliefs and vision of science. Nevertheless, results obtained have been considered to be acceptable explanations of phenomena. *(Participant #7)*

> Absolutely, slowly this course has demonstrated that major scientific discoveries are the result of intuition guided by facts, conjectures and speculations. For example, Einstein and Millikan, among others. *(Participant #12)*

> The veracity of this assertion based on the history of science has been demonstrated by various experiments (e.g., Millikan, Rutherford, Bohr, etc.). *(Participant #17)*

> If it had not been so, I am sure that we could not have made so much progress in science and that we still have to go a long way. *(Participant #18)*

The following is an example of an explanation provided by a participant who partially agreed with this assertion:

> This contrasts with how science is taught these days by following the scientific method, starting from secondary school. *(Participant #13)*

The following response is an example of an explanation by a participant who disagreed with this assertion:

> Yes, there is previous knowledge that helps us to look for new ideas. For example, at present one must read at least 50 articles in order to have an idea

of the topic to be researched. However, this is not based on conjectures and speculations. *(Participant #19)*

It is interesting to note that the number of participants who agreed (Informed view of the NOS) with this assertion was the highest (n = 16), as compared with other assertions. It is plausible to suggest that this may due to one or both of the following facts. (a) This assertion is based on the views of two Nobel laureates (Cooper and Perl) and this may have carried more weight for some participants. Of the seven participants who partially agreed with Assertion #4, five changed their view and agreed on Assertion #5. (b) In Assertion #4, the main idea was based on presuppositions guided by data, theory, and conjectures. In contrast, in Assertion #5, the main idea was based on intuition guided by facts, conjectures, and speculations. It seems that participants considered intuition to be more important than presuppositions.

Conclusions and Educational Implications

Results obtained in this study show that on four (1, 3, 4, and 5) of the five assertions, most participants (60–80%) had an Informed view (Response (a)) of the NOS. Research Assertion #2 (20% had an Informed view) was difficult to understand for most participants, as they considered that an "imposed" vision of the world would go against the scientific community and the libertarian ethos of science. The history of science, however, shows that it is precisely the debate (competitive cross-validation)itself that leads one part of the scientific community to accept an interpretation before a consensus has been achieved. In Cooper's perspective on teaching science this may be considered an "imposed" view of science, similar to the impressionists. The proportion of participants with a Partially Informed view (Response (b)) of the NOS ranged from 15% (Assertion #5) to 60% (Assertion #2). Few participants had a traditional/naive view of the NOS (Assertion #2 being the highest, 15%).

A further analysis was performed to demonstrate the consistency of the results obtained based on participants' responses to the five Research assertions: (1) Seven participants were found to have selected Response (a) on 4 of the 5 assertions. This was considered to be a consistently Informed view of the NOS; (2) 12 participants selected Response (a) on either 2 or 3 of the assertions. This was considered to be a consistent Partially Informed view of the NOS; (3) One participant selected Response (a) on one assertion and Response (b) on four assertions. This was considered to be consistent with a Traditional/Naive view of the NOS.

It is concluded that the views of a Nobel laureate within an HPS perspective, can be helpful in a reflective and explicit, activity-basedcourse to facilitate in-service teachers' understanding of 'science in the making' and, consequently, the NOS. In a recent study, Eshach (2009) has argued that teaching science based on case stories of Nobel Prize laureates can stimulate students' interest.

At this stage a word of caution is necessary lest it be assumed that both the textbooks and the science teacher (both at the school and introductory university level) need only emphasize "cutting-edge"research in the context of Nobel Prize winners. Progress in science is a complex process that is based on various forms of research. For example, there is merit in doing the everyday lab activities to refine results and theories, referred to as "normal science" by Kuhn (1970). However, Kuhn clearly differentiates "normal" and "revolutionary" science, precisely due to the accumulation of anomalies and difficulties with a particular paradigm, that requires a radical restructuring of our knowledge based on "cutting-edge"research. Furthermore, most of our textbooks and science curricula emphasize "normal science" and hence this book, by emphasizing cutting-edgeresearch, establishes an equilibrium that provides a deeper understanding of progress in science.

Based on the results obtained, it would be helpful for teachers to take the following into consideration:

1. It is not the "objective criteria" or "logic of science" that helps us to achieve consensus but, rather, it is the competitive cross-validationbased on peer review that facilitates understanding in science.
2. The peer review process involving interactions among members of the scientific community itself is based on conflicts, controversies, and rivalries.
3. Although, in general the scientific community is the final arbiter in scientific controversies, at times it is difficult to achieve consensus. Consequently, a particular interpretation may be accepted by the community (pending controversy) and thus considered as "imposed" by some.
4. The image of the traditional scientific method is still important for the classroom teacher. Based on experiences with NOS-basedcourses, a new vision of science seems to be emerging.
5. For some teachers there could be a dilemma between the traditional science curriculum (e.g., scientific method) and the new vision of science based on the HPS.
6. In order to be effective, NOS-basedcourses need to be reflective and explicitly based on the topics of the curriculum with which the participants are familiar, namely 'science in the making.'
7. Juxtaposition of experiments with conjectures can facilitate an understanding of the underlying presuppositions and the dynamics of scientific progress.
8. It is difficult for teachers to understand how textbooks can present a vision of science that is systematic and rigorous, whereas scientific practice itself is frequently based on conjectures, presuppositions, and controversies.
9. In some cases it is not easy to reproduce results from original experiments in lab courses. The HPS can provide a framework to interpret lab data and thus understand the dilemma faced in scientific practice.
10. A scientist must have imagination to see beyond what others cannot see, based on intuition guided by facts, conjectures, and speculations.

7

MARTIN PERL'S (NOBEL LAUREATE) PERSPECTIVE ON THE NATURE OF SCIENCE AND TEACHING SCIENCE

Introduction

Martin L. Perl was the recipient of the 1995 Nobel Prize in physics for his discovery of the Tau Lepton, based on a 16-yearhistory (1963–1979), when all experimental measurements agreed with the hypothesis that the Tau was a lepton produced by a known electromagnetic interaction (Perl et al., 1975; Perl, 1997). The objective of this chapter is to present a brief account of the discovery of the Tau Lepton in order to understand the nature of science (NOS) and then draw implications for teaching science.

It is important to note that Martin Perl has also done considerable experimental work on the isolation of elementary particles with fractional electrical charge, quarks (Perl, Lee, & Loomba (2004). Given the difficulties involved in isolating quarks, Perl and Lee (1997) have designed a philosophy of speculative experiments. In order to isolate quarks, Perl and colleagues have used a Millikan style methodology improved by modern technology (image-processingelectronics and high-speed computers). Perl's methodology and the philosophy of speculative experiments also help to understand Millikan's methodology in the oil drop experiment and the Millikan–Ehrenhaft controversy (for details, see Niaz, 2009a, Chapter 13). Perl has now abandoned his active quark research (SLAC-PUB-13512Jan., 2009).

Perl's Discovery of the Tau Lepton

Perl's interest in lepton physics started with his dissatisfaction with the strong-interaction physics of the 1960s and from a desire to discover fundamental facts in particle physics (Perl, 2004, p. 405). This led him to do experiments with the electron and the muon, elementary particles that do not participate in strong interactions but, rather, in the electromagnetic, weak, and gravitational interactions. There were two puzzles about the properties of the electron (*e*) and the muon (μ) in the 1960s: (1) Why is the muon 208 times heavier than the electron, when their properties with respect to particle interaction are the same? and (2) As compared to the electron, why is the muon unstable (lifetime = 2.2×10^{-6} second)? On arriving at the Stanford Linear Accelerator Center (SLAC) in 1963, Perl proposed to look for the unknown differences between the electron and the muon, and speculated that there might be more leptons similar to the electron and the muon, namely unknown heavier charged particles: "I dreamed that if we could find a new lepton, the properties of the new lepton might teach us the secret of the electron–muon puzzle" (Perl, 2004, p. 407). Early experiments using photon beams and later inelastic scattering of muons on protons, did not provide the expected results. This led Perl to conclude:

> *Experimental science is a craft and an art, and part of the art is knowing when to end a fruitless experiment. There is a danger of becoming obsessed with an experiment even if it goes nowhere . . . I avoided obsession and gave up on the scattering experiment. That turned out to be a good decision because modern experiments have shown that these scattering experiments do not illuminate any differences between the electron and the muon beyond the mass difference. (Perl, 2004, pp. 408–409, original italics)*

After the lack of success of the early experiments, Perl considered the possibility of the existence of charged leptons more massive than the electron and muon, called heavy leptons. By the late 1960s the concept of a heavy lepton (*L*) along with its decay modes was the subject of speculation in the research literature. This led Perl to conclude:

> *This brings me to the question I raised in the Introduction: is there more and broader speculation these days in particle-physics theory than forty years ago? Judging by the various ideas of forty years ago about possible types of leptons, we were rather timid about speculations. There was a fear of being thought unsound. There was reluctance to stray too far from what was known. Today the only limit to theoretical speculations about particle physics is that the mathematics be correct and that there be no obvious conflict with measured properties of particles and reactions. One example is string theory with all its different forms and extensions. Another example is the recent, tremendous amount of interweaving of particle physics with astrophysics and cosmology. I think this is a good change. It can stimulate the experimenter to go in new directions, but the experimenter must be cautious as to how she or he uses time. If possibly relevant data have already been produced, then it may be relatively fast to test the speculation against the data. But if a new experiment must be built to test the speculation, that is another story. (Perl, 2004, pp. 409–410, original italics)*

After surmounting difficulties with funding, the construction of the SPEAR electron–positroncollider at SLAC started at the end of the 1960s, along with collaboration with the Lawrence Berkeley Laboratory (LBL) group. The purpose of such detectors (SLAC–LBL)is to detect the type and vector momentum of most of the particles (e.g., electrons, photons, muons, protons, etc.) coming from a reaction taking place at the center of the detector. This led Perl to conclude:

> *These detectors are necessary to obtain the complicated and often subtle data in modern experiments. But large detectors have come at a human cost. It is no longer possible for a few people to build and operate such detectors. Hence there are often hundreds of experimenters in a typical group and the new very large and complicated detectors require groups with more than a thousand members. Of course such detectors produce tremendous amounts of data. (Perl, 2004, pp. 413–414, original italics)*

In 1971, Perl and colleagues at SLAC–LBLsubmitted a research proposal in which the heavy–leptonsearch (Perl's major interest), "was left for the last and allotted just three pages, because to most others it seemed *a remote dream*. But the three pages did contain the essential idea of searching for heavy leptons using electron–muonevents" (Perl, 2004, p. 414, italics added). Perl's colleagues thought that including more about heavy leptons and the electron–muon problem would unbalance the proposal. This did not deter Perl, and he persevered by submitting a 10–page"Supplement" which emphasized that the detector could simultaneously collect relevant data with respect to the following research questions (reproduced in Perl, 2004, p. 415):

1. Are there charged leptons with masses greater than that of the muon?
2. Are there anomalous interactions between the charged leptons and the hadrons?

These questions were, of course based on Perl's basic strategy of studying muon–proton inelastic scattering.

By June 1975, Perl and his group had "good" experimental evidence for electron–muon events, from the SLAC–LBLdetector, which were presented at the Canadian Institute of Particle Physics, McGill University, Montreal. Furthermore, they discussed possible sources of the electron–muon events: heavy leptons, heavy mesons, or intermediate bosons. Of the 126 particle–pairevents reported, at least 24 could be attributed to electron–muon events, which was the strongest evidence at that time for the Tau Lepton. Still, Perl was cautious and concluded, "I was still not able to specify the source of the electron–muonevents: was the source new leptons, new mesons, or new bosons? But I remember feeling strongly that the source was heavy leptons. It would take two more years to prove that" (Perl, 2004, p. 417). This clearly shows Perl's strong belief (presupposition, guiding assumption) with respect to the Tau Lepton.

Based on further experimental evidence and refinement of the procedures, Perl and colleagues finally published in December 1975, their crucial article, which concluded:

> We conclude that the signature *e*–μ events cannot be explained either by the production and decay of any presently known particles or as coming from any of the well–understood interactions which can conventionally lead to an *e* and a μ in the final state. A possible explanation for these events is the production and decay of a pair of new particles, each having a mass in the range of 1.6 to 2.0 GeV/c^2. *(Perl et al., 1975, p. 1492)*

Perl and colleagues, however, were not yet prepared to claim that they had found a new charged lepton, and in order to accentuate their uncertainty they denoted the new particle by U for "unknown" in some of their 1975–1977 papers. Despite the uncertainty, and the need to convince the world that the electron–muon events were significant (caused by the decay of a pair of heavy leptons), Perl was still optimistic, "Thus in 1975, twelve years after we began our lepton–physicsstudies at SLAC, these studies bore fruit" (Perl, 2004, p. 418). Besides his research program, guiding assumptions, new technology (electron–positron collider), Perl attributed the success to:

> *I had smart, resourceful, and patient research companions. I think these are the elements that should be present in* <u>*speculative experimental work:*</u> *a broad general plan, specific research methods, new technology, and first-class research companions. Of course, the element of luck will in the end be dominant. (Perl, 2004, pp. 418–419, original italics, underline added)*

Despite publication of the experimental results (Perl et al., 1975), there was still confusion and uncertainty about the validity of the data and its interpretation. One of the impediments to the acceptance of the tau as the third charged lepton was that no other laboratory had reported similar results. Looking back over those years, Perl (2004) reported: "It was a difficult time. Rumors kept arriving of definitive evidence against the tau: electron–muonevents not seen, the expected decay modes not seen, theoretical problems with momentum spectra or angular distribution" (pp. 420–421). In 1976, however, things started to change when outside experimental confirmation for anomalous muon events was reported by Cavalli–Sforzaet al. (1976). The second confirmation came from Perl's own group at the SLAC–LBLdetector, providing support for the decay of a heavy lep–ton (Feldman & Perl, 1977). The third confirmation came from an experiment at the DORIS electron–positron storage ring in Hamburg (PLUTO–Collaboration, et al., 1977). These events convinced Perl that he was right about the existence of the tau as a sequential heavy lepton.

If the tau was indeed a sequential heavy lepton, two substantial semi-leptonic decay modes had to exist. One of the problems in finding these decay modes (during the years 1977–1979) was the poor efficiency for photon detection in the early detectors. To resolve this problem new detectors (such as Mark II) were put into operation at SLAC–LBL facilities. Given this new efficiency of the detectors, the decay mode was detected and measured by various groups of researchers in different parts of the world, including Perl's own group (Abrams & Perl, 1979).

Finally, after all the difficulties that involved improved experimental equipment and interpretations, the stage was set for Perl (2004) to conclude:

> By the end of 1979, all confirmed measurements agreed with the hypothesis that the tau is a lepton produced by a known electromagnetic interaction and that, at least in its main modes, it decays through the conventional weak interaction. So ends the sixteen year history, 1963 to 1979, of the discovery of the tau lepton and the verification of that discovery. *(p. 424)*

Interestingly, just as Perl's discovery of the Tau Lepton solved the puzzles related to the nature of the electron and the muon, it created additional puzzles with respect to the Tau itself. Perl (2004) considers this as an irony of his discovery and suggested the following as a possible explanation: "The discovery of the last decade that one type of neutrino can change into another type of neutrino may lead eventually to an understanding of what constitutes the intrinsic nature of elementary particles. But we are not there yet" (p. 424). This shows how the resolution of one puzzle leads to another and, consequently, the need for further research. Kragh (1999) has referred to the additional puzzles produced by Perl's discovery of the Tau in the following terms: "The accompanying tau neutrino was assumed, rather than detected. The acceptance of the tau lepton (or tauon) implied that the symmetry between quarks and leptons—or between strong and weak interactions—was no longer satisfied. Two new quark flavors were needed and, as we shall see shortly, they were later found" (p. 343).

Perl's Research Program: 'Science in the Making'

Based on the historical reconstruction of Martin Perl's work on elementary particle physics it is plausible to elaborate the following aspects of his research program, namely 'science in the making':

Perl's theoretical framework (presuppositions): Perl's interest in weak interactions of elementary particles led him to consider two puzzles with respect to the electron (*e*) and the muon (μ). Both the electron and the muon have the same properties with respect to weak particle interactions. However, the controversial and conflicting aspect consisted of the fact that compared to the electron, the muon was unstable and 208 times heavier. To understand the difference between the muon and the electron, Perl speculated (even dreamed) that other leptons (heavy charge particles) existed.

Research questions: After having recognized the controversial nature of the properties of the electron and the muon, Perl formulated the following research questions: (a) Are there charged leptons with masses greater than that of the muon? and (b) Are there anomalous interactions between the charged leptons and the hadrons?

Operationalizing heuristic principles: In order to specify the source of the electron–muon events, Perl considered the following as possible candidates: new leptons, new mesons, or new bosons. At the same time he strongly felt that the source was heavy leptons.

Designing experiments: Perl was actively engaged in the design and execution of the experiments even while he was at Michigan and much more so after his arrival at SLAC in 1963. Early experiments using photon beams and, later, inelastic scattering of muons on protons, did not provide the expected results. Perl avoided obsession and gave up on scattering experiments. This led to the construction of the SPEAR electron–positron collider at the end of the 1960s, which involved Lawrence Berkeley Laboratory (LBL) group and SLAC and led to the SLAC–LBL research proposal. The new facility still lacked efficiency for photon detection and thus a new detector (Mark II) was built which helped resolve the decay mode problem of the heavy lepton during 1977–1979. Actually, during the period 1963–1979, Perl and other researchers were constantly engaged in improving experimental equipment based on their research questions.

Understanding the nature of science: Historical reconstruction of the discovery of the Tau Lepton by Martin Perl, provides an insight into 'science in the making' and eventually the NOS, by highlighting the following aspects:

1. Experimental strategies can lead to dead ends and a scientist needs to be aware when to look for new ones. For example, Perl (2004, pp. 408–409) avoided being obsessed with his early scattering experiments.
2. Speculations in particle physics and scientific research in general can stimulate the experimenter to look for new alternatives. The only limit to this is that there is no obvious conflict with known experimental findings. Perl (2004, p. 409) considers string theory to be an example of theoretical speculations in particle physics. Interestingly, Perl considers that 40 years ago physicists were rather timid with respect to speculations, one example being the possible types of leptons.
3. Experimental particle physics now depends on large detectors, which produce tremendous amounts of data (Perl, 2004, pp. 213–214). This makes

interpretation of data extremely difficult and requires the collaboration of large groups of scientists.

4. Scientists need to persevere with their theoretical framework (presuppositions) before abandoning it and looking for new frameworks. Perl (2004, pp. 413–414) relates that when the SLAC–LBL proposal was submitted in 1971, it devoted only three pages to the search for heavy leptons to explain the difference between the electron and the muon, as most of his colleagues considered it to be "a remote dream" that would "unbalance the proposal." Under these circumstances Perl persevered (as he strongly believed in his presuppositions) by including a 10–pagesupplement in order to "balance" the proposal.
5. Experiments are generally difficult to implement and in experimental particle physics even more so. However, interpretation of experimental data is much more difficult. Perl et al. (1975) published their crucial data, but due to the uncertainties in interpretation were not prepared to claim that electron–muon events came from the decay of a pair of heavy leptons and, consequently, for two years (1975–77) the new particle was denoted by U for "unknown."
6. Speculative experimental work in particle physics and research in general is facilitated by the following elements (Perl, 2004, pp. 418–419): A broad general plan, specific research methods, new technology, capable research companions, and some degree of luck.
7. Additional puzzles: Perl (2004, p. 424) considers it ironic that his discovery of the tau provided answers to the puzzles with respect to the differences between the electron and the muon. However, now the electron–muon puzzle has expanded into the electron–muon–tau puzzle. This illustrates the progressive nature of science, namely the resolution of one set of problems can lead to the identification of new problems that in turn need resolution.
8. A constant interaction between Perl's presuppositions and experimental verification of the Tau Lepton over a period of almost 16 years, which clearly shows the importance of both.

Perl's Perspective on Teaching Science

On receiving the Nobel Prize for Physics in 1995, Martin Perl started his acceptance speech with the following words: "My first thoughts in writing this lecture are about the young women and young men who are beginning their lives in science: students and those beginning scientific research" (Perl, 1997, p. 168). Indeed, Perl's research experience extending over 50 years can have important implications for teaching science. Next, in his speech he referred to the human vicissitudes involved in doing experimental work:

a. Pleasure, when an experiment is completed and the data safely recorded.
b. Anxiety, when an experiment does not work well or breaks.

c. Pain, when an experiment fails or when an experimenter does something stupid.
d. In the discovery of the Tau Lepton, the ups and downs of his emotions extended over years.
e. Stay away from lines of research where many competent people are working. Interestingly, when Perl received his Ph.D. in 1955, he had three job offers: Physics Departments at Yale, University of Illinois, and the University of Michigan. What did Perl (1995) do? "At that time, the first two Physics Departments had better reputations in elementary particle physics, and so I deliberately went to Michigan."
f. There is nothing wrong with an experimental idea as long as you are the first to use it.
g. For a speculative experimenter it is good to have two experiments going, one in operation and the other being built.
h. Avoid obsession with a particular experiment.
i. In order to understand physics, look for simplicity.
j. Good speculative experimental work requires: a broad general plan, specific research methods, new technology, first-classresearch companions, and some degree of luck.
k. Interpreting experimental data is difficult and leads to uncertainties.
l. In order to get original ideas on a subject it is good to do experimental work.

The Role of Creativity

Perl (2007) has emphasized the need for developing creativity for science and engineering education:

> Creativity is a broad aspect of humanity, but in engineering and science has constraints that do not exist in other creative activities such as music and art. An improvement of computer architecture, a discovery of a new medicine, a new understanding of the behavior of black holes, an improvement in gasoline engine efficiency—all are creative feats that are nonetheless limited by the laws of nature. We may be wrong about a law of nature. For example the consensus that energy producing, cold fusion is not possible is based upon our present understanding of thermodynamics and nuclear physics. The present understanding may be wrong, but that has to be demonstrated by consistent, quantitative experimentation. *(p. 2)*

This statement has some important implications: (a) Creativity can manifest in various fields of knowledge; (b) Creative feats are constrained by the laws of nature; and (c) Laws of nature may be wrong and in turn may be replaced. This summarizes very succinctly various discoveries in science and engineering.

However, Perl (2007) cautions against considering "Kekulé's dream to discover the ring structure of benzene," as an act of creativity. Next Perl outlines some basic abilities and skills necessary for creativity in science and engineering:

a. Competency in mathematics.
b. Imagination, subject to the constraints of known physical laws.
c. Visualization based on drawing pictures, rough calculations, computer simu–lations, and models.
d. Hands–onand laboratory skills are essential for some types of creative work. However, if someone lacks lab skills she/he can still be creative through design/theoretical work. In this context, Perl (2007) provides interesting anecdotal information with respect to Isidore Rabi (Nobel laureate in Phys–ics): "My Ph.D. thesis advisor, Isidore Rabi, was given the Nobel Prize for his experimental work, but he had few hands–onskills. His graduate students were afraid to let him get close to their apparatus. When he came to the laboratory we immediately engaged him in conversation at the door, hoping he would get bored and leave. In spite of his hands–onlimitations, he had a deep, mysterious way of understanding and visualizing experimental work" (p. 3).
e. Patience and fortitude in looking for good ideas. A major problem with finding good ideas is that it is difficult to separate them from bad ideas. This problem is compounded as bad ideas are also found in the work of famous scientists and due to hero worship these ideas are not mentioned in their biographies. Perl cites the example of Einstein who had many erroneous ideas on unified field theory after 1925. Perl cites the example of Isaacson's (2007), 500–pagebook on Einstein, which devotes only a few pages to ideas that did not work.
f. Freedom to try out new ideas. Most undergraduate and graduate school pro–grams are overloaded with course and lab work. Perl (2007) is critical of such curriculum practice as it does not motivate the students to do creative work: "There is little time for the student to play with ideas, to dream about discoveries and inventions" (p. 5).

How many of our students would be willing to follow Perl's advice on cutting–edge experimental work? Needless to say, Perl clearly brings to the forefront the idea that 'science in the making' is a human enterprise.

8

CONCLUSION

From 'Science in the Making' to Understanding the Nature of Science

An Overview

Different chapters of this book have presented 'science in the making' as a rich landscape in which scientists have to face criticisms, rivalries, conflicts, passionate debates, and alternative interpretations of data by fellow scientists. On the contrary, most of our students, textbooks, and even teachers consider science to be a repository of facts, laws, and theories that need to be memorized. Furthermore, most science curricula promote a simple empiricist view of scientific development that is driven by the traditional scientific method. It seems as if the alternative of teaching science as practiced by scientists would threaten the very structure and the walls of science (Niaz, 2010a, b; Niaz, 2011).

A reconstruction of various historical episodes shows the importance of 'science in the making' through heuristic principles that facilitate our understanding of some important aspects of the nature of science (NOS) (Chapter 2): Experimental data do not necessarily and in a straightforward manner lead to the formulation of scientific theories and laws; A theoretical framework based on a scientist's presuppositions has some speculative aspects; Alternative interpretations of experimental data are controversial and, at the same time, provide a greater understanding of scientific progress; The tentative nature of scientific theories; Scientists often proceed by speculating, hypothesizing, proposing models, offering explanations and formulating a theory, which contrasts with the traditional view of the scientific method; Scientists are in general reluctant to give up and change their presuppositions (hard-coreof beliefs); Postulation and acceptance of a new model is difficult due to the rivalry with existing models; Conjugation of *experiments* and *conjectures*, namely the heuristic principles help scientists to understand and conceptualize experimental data; and Postulation of new hypothetical ideas for which there is no existing experimental evidence is a perfectly acceptable research strategy. It is important to note that these aspects of the NOS are based on a historical reconstruction of actual scientific practice.

A review of the literature in science education reveals that research has focused more attention on identifying students' and teachers' epistemological beliefs (64%), whereas less work (36%) has been done on actual teaching strategies for facilitating understanding of the history and philosophy of science (HPS) (Chapter 3). This shows the need for future research to emphasize the development of new explicit and reflective teaching strategies, based on different topics of the science curriculum. The tentative nature of scientific knowledge continues to be one of the most difficult NOS aspects in most parts of the world. Argumentation in the classroom could provide a congenial theoretical framework for facilitating the NOS based on a historical perspective. Teaching science within a historical perspective leads to the dilemma (for some) of teaching false science (Bohr's model, Newtonian mechanics). This can easily be circumvented by suggesting to students that scientific models/theories are not right/wrong but provide us with a better understanding of the world with respect to a previous model/theory. This in a sense also facilitates a better understanding of how and why scientific models/theories are tentative. There is some evidence to show that the science curricula themselves, by emphasizing certainty and objectivity, may be acting as a barrier to the understanding of the NOS. In contrast to the traditional science curricula which invariably emphasize "science for scientists," we need to ask, "Are we teaching science as practiced by scientists?" An answer to this question may help us to find new and motivating teaching strategies for enhancing students' interest in learning science.

Chapter 4 provides an example of such a teaching strategy for introducing the NOS in the classroom. Based on the results obtained in this study participants acquired considerable experience with respect to: (a) The complexity of the issues involved in 'science in the making'; (b) Teaching the NOS seemed to contradict the traditional science curriculum based on positivism; (c) The tentative nature of scientific theories was considered to be the most important NOS aspect; (d) The theory-laden nature of observations facilitated a scientist's guiding assumptions that played the role of a heuristic principle; (e) Scientific theories do not necessarily emerge in a logical manner; (f) The dual role of a hypothesis in foreseeing the results and providing a tentative explanation; (g) The most important aspect in the development of a theory is the heuristic principle and not the experimental data; (h) Elaboration of models (based on facts and concepts) requires considerable conceptual complexity; and (i) Scientists generally do not follow the scientific method. It is concluded that introduction of the NOS in the classroom requires considerable intellectual effort and conceptual understanding of 'science in the making' based on an HPS perspective.

Next, the role of heuristic principles in understanding the NOS was explored (Chapter 5). Based on the results obtained in this study it is plausible to suggest that teachers acquired considerable experience with respect to: (a) The inclusion of the NOS based on an HPS perspective in contrast to the empiricist perspective of most textbooks and classroom environments; (b) The importance of "heuristic principles" in order to understand different aspects of NOS, as part of the historical episodes; (c) The Lakatosian perspective that considering students' alternative conceptions and its criticism (based on the new ideas) as rival or competing models can facilitate conceptual change. History of science shows that scientists themselves generally experience a similar process in which rival theories can co-exist and explain the same set of data, until a consensus is achieved through peer evaluation; (d) In order to make science objective, textbooks follow the "experimenticist fallacy" which leads science content to be presented as a "rhetoric of conclusions." On the contrary, objectivity in science is a consequence of a social process that requires interpretation of experimental data based on rival sets of "heuristic principles"; (e) Most textbooks conceptualize objectivity as part of the "nature of facts," that is, "truths are out there to be discovered." It would be more motivating for students and teachers if science is presented as a human enterprise in which the difficulties involved in constructing theories are highlighted; (f) The need for creating a new *scenario* in the classroom in which students and teachers could present arguments and counter-argumentsbased on historical reconstructions of various episodes in the history of science.

In order to familiarize students with the views of a practicing scientist (Leon Cooper, Nobel laureate, Physics) with respect to how scientists practice science, was the subject of the next study (Chapter 6). Results obtained in this study showed that on four (1, 3, 4, and 5) of the five assertions based on Cooper's historical perspective, most participants (60–80%) had an Informed view of the NOS. Research Assertion #2 was difficult for most participants and only 25% had an Informed view of the NOS. In this assertion Cooper dealt with an "imposed" vision of science, especially when consensus is difficult to achieve. Some participants considered that an "imposed" vision of the world would go against the scientific community and the libertarian ethos of science. The proportion of participants with a Partially Informed view of the NOS ranged from 15% (Assertion #5) to 60% (Assertion #2). Few participants had a Traditional/Naive view of the NOS (Assertion #2 being the highest, 15%). It is concluded that the views of a Nobel laureate within an HPS perspective, can be helpful in a reflective and explicit, activity-basedcourse to facilitate in-serviceteachers' understanding of the NOS.

The experience of a practicing experimental scientist (Martin Perl, Nobel laureate, Physics) provided a glimpse of cutting-edgeexperimental research, based on the discovery of the Tau Lepton (Chapter 7). It is important to highlight the following aspects of Perl's 16-yearstruggle (1963–1979): (a) Experimental strategies can lead to dead ends and a scientist needs to be aware as to when a change is necessary; (b) Speculations can stimulate the experimenter to look for new alternatives; (c) Scientists need to persevere with their theoretical framework (presuppositions) before abandoning the research; (d) Experiments are difficult to implement and even more difficult to interpret. The Tau Lepton was initially denoted as U for "unknown," as no satisfactory interpretation was available, that could be acceptable to the scientific community; (e) Speculative experimental philosophy is facilitated by the following elements: a broad general plan, specific research methods, new technology, capable research companions, and some degree of luck; (f) Scientific research is accompanied by pleasure, pain, anxiety, and ups and downs in emotions over extended periods of time. Perl's experience clearly shows how scientific research constitutes a human enterprise that involves hopes, dreams, aspirations, and frustrations.

In Chapter 1, I outlined seven important aspects based on questions that in my opinion facilitate an understanding of the NOS. In the remaining sections of this chapter I shall provide plausible answers to these NOS–orientedquestions based on research presented in various chapters of this book.

Why Is it Important to Understand How Science Works?

In most parts of the world science is presented as a finished product with no reference to the difficulties faced by the scientists in doing experiments and interpreting experimental data. Duschl (1990) has referred to this as the "final form science" and argued cogently for presenting to students a more human and plausible glimpse of how science works:

> To expect learners to accept at face value the credibility of scientific data without any clues concerning the process/tools of data collection, data manipulation, and data interpretation is wrong. Such an expectation once again reinforces *final form science* and presents science as a series of success stories. *(p. 89, emphasis added)*

Indeed, all "success stories" refer to major scientific discoveries, which are inevitably preceded by controversies and alternative interpretations of data. Various chapters of this book have presented a picture of how science works, namely the NOS and 'science in the making':

a. Members of the scientific community doing actual research are in a better situation to understand the various aspects of the NOS. Nevertheless, Chapter 3 showed that in some cases even these scientists faced a tension between how and what part of the research experience can be published (Schwartz & Lederman, 2008; Yore et al., 2004). In general, scientists tend to avoid publishing the creative, human, and social aspects of their experience. It is precisely these aspects that can facilitate a better understanding of 'science in the making.'
b. Chapter 4, after having discussed various aspects of the NOS, asked in-service teachers to select four aspects of the NOS that in their opinion were

the most important. Later in the course the participants were asked to select the single most important aspect of the NOS. Interestingly, on both questions a majority of the participants selected the same three aspects, in decreasing importance: The tentative nature of scientific theories; that observations are theory-laden,and the role of conflicts, controversies, and rivalries in scientific progress. These three aspects provide a fairly good picture of how science works. One of the participants expressed her/his understanding of the tentative nature of scientific knowledge (according to de Berg, 2010 "discipline in transit") in the following terms:

> This NOS aspect helps us to understand that the "definitive" in science has not been found and hence we have to keep investigating so that we and our students can also make some contribution toward scientific progress . . . This will make students understand that in science we shall always be looking for the "truth" or perhaps "half truth."

This clearly shows that classroom instruction can facilitate students' and teachers' understanding of how science works.

c. Scientists generally offer resistance to changes in their "hard-core"of beliefs (Lakatos, 1970). Similarly, if our students do not, "resist conceptual change and easily accept all new ideas, then [their] cognitive structures would be in a chaos and it would be difficult to establish a firm base for problem solving" (reproduced from a science teacher, Chapter 5).
d. In Chapter 6, one of the science teachers presented the following perspective with respect to how science works:

> Situating in the historical context we find that scientists have always developed new concepts, some of which were controversial in their epoch and caused disagreements and rejections. However, with the passage of time other scientists come to accept theories that were previously rejected.

The "historical context" provides various examples of how an "imposed" vision (unfamiliar or not fully developed) is slowly adopted by the scientific community, e.g.: Rutherford's nuclear atom, Millikan's interpretation of the oil drop experiment, Einstein's hypothesis of lightquanta.

e. Leon Cooper has juxtaposed *experiments* with *conjectures*. In Chapter 6, one of the science teachers expressed this idea in the following terms: "Our present day textbooks project science as a rigorous and systematic demonstration of how the events occurred, and not as a result of the conjectures and presuppositions of the scientists who devoted their lives to science."
f. In Chapter 6, based on the teaching perspectives of Leon Cooper and Martin Perl (Nobel laureates), one science teacher provided the following insight

with respect to how science works, "In the case of cutting-edge research, based on previous findings a scientist can have an intuition with respect to future results. Consequently, the scientist can help to make the field move forward by speculating and making conjectures."

g. According to Perl (2004), speculations in particle physics and scientific research in general can stimulate the experimenter to look for new alternatives (Chapter 7). The only limit to this is that there is no obvious conflict with known experimental findings.
h. Most textbooks and science curricula make students believe that scientists are involved in straightforward, matter of fact laboratory activities, leading to new discoveries. Perl (1997) in his Nobel Prize acceptance speech, based on his personal experience, referred to how he had to face pleasure, anxiety, pain, and ups and downs in emotions, extended over many years (Chapter 7).

Is it Not Sufficient for Students to Learn the Content of Science?

It is plausible to suggest that if students learn only the content of science they would be deprived of the rich, human, intriguing, and complex side of science. It is precisely these personal struggles of a scientist centered within a scientific community that can motivate students to pursue science as a career.

In Chapter 4, in-service teachers (graduate students) were asked, "Do you think there is a science that we do not teach?" All participants agreed that there is, based on the premise that we teach science content and not the NOS. Participants not only agreed that the way we teach science leaves out important aspects of scientific development but also provided examples from the history of science that go beyond a simple presentation of science content. Some participants suggested that the inclusion of the following issues could be helpful and motivating for students: Where does an original idea come from? Why do scientists have to publish their work? Is competition between scientists inevitable? Who finances research? and How do we recognize the merit of a scientist? These issues can perhaps be included at the end of a chapter in sections such as: Novel aspects of science, New tendencies in research and education, Science for the future. Furthermore, participants were critical of their undergraduate science education in which both the science curriculum and textbooks endorse the view that the success of famous scientists depends primarily on their following the scientific method strictly.

In Chapter 5, one of the science teachers argued cogently as to why it is not sufficient to learn the content of science, and the need for going beyond:

> Classroom discussions must emphasize that when two or more theories clash based on basic principles and arguments, these can co-exist for some time and one is not necessarily replaced by the other. This will show to students that we do not have an absolute "truth" and that scientific theories change.

Do Students Have to Learn How and Why a Scientist Performed an Experiment?

It is generally believed that conducting experiments and the ensuing accumulation of experimental data unambiguously leads to the formulation of scientific theories and laws (cf. Chinn & Malhotra, 2002; Cushing, 1989). The following examples from the history of science and classroom practice provide a better understanding:

a. Astronomers (Leverrier, Adams, & Galle) started looking for the planet Neptune after it was hypothesized that it was causing a gravitational effect leading to irregularities in the orbit of another planet, Saturn (for details, see Chapter 1).
b. Detection of the elementary particle neutrino was started by scientists (Reines & Cowan) after its role was hypothesized in order to uphold the law of conservation of energy in beta decay reactions (for details, see Chapter 1).
c. Millikan did not "design" the oil drop experiment but, rather, "discovered" it. In other words, Millikan had a prior belief (presupposition) that the elementary electrical charge existed and this led him to the need for experimental determination (Chapter 2).
d. It is suggested that the traditional understanding of theories and laws as representing absolute "truth" does not provide an epistemic environment conducive towards an understanding of the role of experiments in theory construction (Chapter 3, Grandy & Duschl, 2007; Liu & Tsai, 2008).
e. In Chapter 4, in-servicescience teachers (graduate students) were asked to agree/disagree with the following affirmation: "Accumulation of experimental data is the most important aspect in the development of a theory/law." All 16 participants disagreed and the following is an example of participants' thinking: "I am in total disagreement as a simple accumulation of data conduces to nothing. The most important aspect in the development of a theory is the 'heuristic principle' (Schwab, 1962)."

Does it Help Students to Know that the Same Experimental Data was Interpreted Differently by Another Scientist?

The following examples from the history of science illustrate how different interpretations of the same experimental data lead to conflicts and controversies:

a. E. Rutherford (1911) presented his model of the nuclear atom based on data from alpha particle experiments. J.J. Thomson, renowned world-masterin elaborating atomic models did the same experiments in his lab and rejected Rutherford's nuclear atom. Rutherford's model was based on the hypothesis of *single scattering* of alpha particles. In contrast, Thomson interpreted the same data based on the hypothesis of *compound scattering* (for details, see Chapter 2). A bitter dispute ensued between Thomson and Rutherford that lasted for many years.

b. R. Millikan and F. Ehrenhaft, interpreted very similar experimental data to formulate the universal charged particle "electron" and particles with fractional charge, "sub–electron,"respectively. The controversy lasted for many years and for some time the scientific community itself could not provide feedback (for details, see Chapter 2).
c. In Chapter 4, in–servicescience teachers (graduate students) were asked to agree/disagree with the following affirmation: "Based on experimental data, scientific theories emerge in a logical manner." All participants except one disagreed with this affirmation and some referred to the two examples provided above (Thomson/Rutherford and Millikan/Ehrenhaft). One of the participants argued succinctly, "I disagree as it is not the experimental data but the scientific community that in the final analysis validates scientific theories." Indeed, the role played by the scientific community in the two examples cited above was crucial (cf. Niaz, 2009a).
d. In Chapter 6, one of the science teachers observed, "[T]his happens very frequently in science, and forms part of the nature of science. For example, the controversy between Thomson and Rutherford (cf. point (a) above)—same experimental results and different interpretations."
e. Martin Perl's discovery of the Tau Lepton shows the difficulties and uncertainties involved in interpreting experimental data (Chapter 7). Even after having the crucial data in 1975, Perl could not convince the scientific community that he had discovered the new particle, which was in the interim denoted by U for "unknown." It took two more years (1975–1977) to finally convince the scientific community.

Do We Present a False Image of Science in our Textbooks and Classrooms?

If we ignore the following aspects of the NOS, we shall be presenting a false image of science to our students:

a. Experimental data do not necessarily and in a straightforward manner lead to progress in science. Examples: Law of multiple proportions in chemistry; Millikan's and Ehrenhaft's interpretations of data to determine the elementary electrical charge; Davisson and Germer's interpretation of diffraction of electrons by crystals (for details, see Chapter 2).
b. The theoretical framework of a scientist (presuppositions) can include some speculative aspects. Examples: Maxwell's kinetic theory of gases; Bohr's model of the atom; de Broglie's postulation of wave–particle duality (for details, see Chapter 2).

c. The tentative nature of scientific theories. Examples: Early attempts to understand the periodic table of chemical elements; Atomic models of Thomson, Rutherford, Bohr, Bohr–Sommerfeld, and the Wave mechanical model (for details, see Chapter 2).
d. Alternative interpretations of experimental data lead to controversies. Examples: Accommodation and prediction in the development of the periodic table; Periodicity in the periodic table as a function of an inductive generalization or atomic theory; Lenard's trigger hypothesis, Planck's quantum analog and Einstein's lightquanta to explain the photoelectric effect; Millikan's postulation of the electron and Ehrenhaft's sub–electron(for details, see Chapter 2).
e. Scientists are generally reluctant to change their presuppositions. Examples: Thomson's reluctance to give up his own model of the atom and accept Rutherford's nuclear atom; In the Millikan–Ehrenhaft controversy, both scientists strongly believed in their presuppositions and consequently in their interpretations (it is generally ignored that Ehrenhaft adhered to anti–atomism based on the work of Mach and Ostwald); Millikan's reluctance to give up the classical wave theory of light and accept Einstein's lightquanta.
f. Postulation and acceptance of a new model is difficult due to the rivalry with existing models. Examples: Difficulties and uncertainty faced by Rutherford (he delayed the publication of his model for almost two years) before postulating his nuclear atom; de Broglie's postulation of wave–particle duality.
g. Observations are theory–laden,that is, there are no unbiased observations as suggested by the scientific method. Examples: Varying interpretations of the photoelectric effect based on the theoretical frameworks of Lenard, Planck, Millikan, and Einstein; Millikan's interpretation of data from his oil drop experiment (for details, see Chapter 2).
h. In most parts of the world the science curriculum includes the atomic models of Thomson, Rutherford, and Bohr. These models are not valid any more or, in other words, constitute "false science." Chapter 3 showed that teaching "false science" in this context helps to illustrate how science progresses and thus to understand the tentative nature of scientific knowledge.
i. In Chapter 5 one of the science teachers referred to HPS, inductivism, and objectivity in the following terms: "In contrast to the HPS perspective, the inductivist vision is rigid and does not contemplate 'transgressions' of objectivity, precision and methodology. This rigidity makes any change in a paradigm difficult, and textbooks continue to repeat the same framework." Transgression of objectivity is an important concept and leads to a dilemma: If scientists are absolutely "objective" then the path from data to theory would be free from controversy. Philosophers of science have referred to this as a "paradoxical dissociation": "While nobody would deny that *science in the making* has been replete with controversies, the same people often depict its essence or end product as free from disputes, as the uncontroversial rational

human endeavor par excellence" (Machamer, Pera, & Baltas, 2000, p. 3, emphasis added). Presenting science to our students as free from controversies is clearly a false image of science.

j. One of the science teachers referred to the relationship between experiments, observations, and objective knowledge in Chapter 5: "Different aspects of the nature of science (including controversies and idealization) constitute the emerging vision. However, in many parts of the world positivism still wields considerable influence, according to which 'experiment and observation' are the hallmark of 'scientifically proved objective knowledge'." Experiment as the hallmark of science does not facilitate the emerging vision and thus presents a false image of science.
k. One of the science teachers referred to the difficulties involved in teaching about the nature of objectivity as most textbooks and curricula consider that, "objectivity is 'blind' and does not depend on the observer but rather is part of the nature of facts" (Chapter 5). Indeed, it is difficult to explain to the students that Einstein went beyond Newton, not because Newton's work lacked objectivity, or was "wrong." It is precisely the discussion of such issues that can facilitate an understanding of the NOS.
l. Our notions of good science are based on the universal scientific method, "which is more of a recipe and 'Lord's Prayer' charged with empirico-positivist notions" (reproduced from a science teacher, Chapter 5). This contrasts sharply with 'science in the making' based on actual historical episodes.
m. Students and even teachers generally believe that what the scientist knows is based on the experiment. Based on Leon Cooper's perspective of teaching science (Chapter 6), one science teacher provided the following insight, "To be sure, it would be impossible to design an experiment if we do not know what we want to know. It is precisely in this context that presuppositions play an important role."
n. Scientists generally persevere with their theoretical framework (presuppositions) before abandoning it and looking for new frameworks. Martin Perl persevered in his search for heavy leptons, even though some of his colleagues considered it to be "a remote dream" (Chapter 7).
o. It is generally believed that famous scientists have good ideas that facilitate new discoveries. Perl (2007) has explained that it is good to separate good ideas from bad ones, especially due to the fact that even famous scientists have bad ideas. Consequently, patience and fortitude are required in looking for good ideas (Chapter 7).

Is the False Image of Science Conducive towards a Better Understanding of Science?

In Chapter 4, one of the science teachers summarized 'science in the making' in the following terms: "The role of science is to create, revise and refute theoretical models. This shows the changing or tentative nature of scientific development [and] that teaching science should be an investigative process." It is plausible to suggest that any science curriculum or textbook that departs significantly from this conceptualization would be presenting a false image of science and hence would not be conducive towards a better understanding of science. A false image of science would inevitably emphasize science content (cf. Duschl, 1990, final form science) leaving out the personal struggle of a scientist to convince the scientific community. Of course, a science textbook cannot assume the role of presenting the history of science. Nevertheless, some discussion of the origin of ideas and the difficulties involved can effectively engage students and enhance motivation.

One of the science teachers referred to the relationship between "truth," rhetoric, and memorization in the following terms: "As teachers we generally present the opinion of a famous scientist as authoritative and the absolute 'truth'. This leads the students to memorize the 'truth' and ultimately leads to the traditional 'rhetoric' of science" (Chapter 5). All of us are aware of how science textbooks present famous scientists such as Thomson, Rutherford, Bohr, Millikan, Planck, and Einstein as "heroes." This perspective distorts the historical reconstructions in which these scientists had to face the critical evaluations of their peers. A false image of scientists as infallible leads students to accept ideas uncritically and to memorize facts and, consequently, does not facilitate understanding of science.

In Chapter 5, one of the science teachers referred to the relationship between the tentative nature of scientific theories and memorization: "Based on my educational background I never thought that scientific laws and theories are tentative and could change and consequently I ended up memorizing them." No wonder our students are more prone to thinking that if science delivers eternal truths, then memorization is the best learning strategy.

Students and teachers generally consider scientific discoveries to be an end in themselves. On the contrary, the history of science shows that, in general, one theory/model is a stepping-stonetowards a new discovery. Martin Perl solved the electron-muon puzzle and in the process created the electron-muon-tau puzzle (Chapter 7). This shows the progressive nature of science, namely, resolution of one set of problems can lead to the identification of new ones. Indeed, students as future scientists can become motivated to adopt science as a career if they know that resolution of one problem leads to another. Furthermore, many students must have pondered and wondered as to what they could contribute towards changing scientific theories.

Does the Science Curriculum Motivate Students to Engage Creatively and Form Part of a Responsible Citizenry?

Motivating students to study science in order to make contributions towards the resolution of social problems can lead to a responsible citizenry. Based on different chapters of this book following aspects can be helpful:

a. The tentative nature of scientific knowledge is generally ignored by the science curriculum. In Chapter 3, we heard that Sadler et al. (2004) have argued that scientists would not devote their lives to the pursuit of knowledge if they had no chance of changing existing knowledge. In other words, emphasizing the tentativeness of science in the classroom can motivate students (just like the scientists) to adopt science as a career and thus contribute towards progress in scientific knowledge.
b. In Chapter 5, one of the science teachers expressed the dilemma between using the actual textbooks (based on the traditional science curriculum) and the motivation required for going beyond: "Like my colleagues, I used to think that our actual textbooks were adequate and that this was the only way of writing them . . . I am now motivated to do further research in order to facilitate conceptual understanding of my students." Another teacher referred to the distinction between understanding science and the "rhetoric of conclusions": "From this perspective [HPS] it is much more motivating to understand science, and especially that it is not a simple rhetoric of conclusions (Schwab, 1974)." The need for a science curriculum that motivates students to engage creatively was highlighted by another teacher:

   > This semester in my 9th grade class, while teaching atomic structure, instead of including many experimental details, I introduced some historical description of the controversy between J.J. Thomson and E. Rutherford. This was so motivating for the students that in their next experiment they wanted to know the origin of the experiments and if there were contradictions between the scientists and how were these resolved.

c. One of the science teachers referred to the difficulties involved in changing the curriculum: "Our society is resistant to change and even the existing curricular structure of our university will not be receptive to the 'new science' [HPS] based on different aspects of the nature of science" (Chapter 5).
d. One of the science teachers expressed his motivation to do research and present the "other side of science":

   > which is not presented in the majority of the textbooks . . . This provides me an opportunity to convince my students with arguments (and their counter-arguments),that we have to go beyond the application of formulae and understand science within a social and historical context that facilitated the construction of these concepts. (Chapter 5)

e. Leon Cooper has compared a scientist to a painter, "The impressionists, for example, were accused of *not being able to see things as they are*. But, having imposed their way of viewing—their vision of the world—it has become a cliché now to see things as the impressionists did. Consider, for example, many magazine ads" (reproduced in Niaz et al., 2010b, p. 48, italics added).

Not to see things as they are means that we can be misled by some observations and hence require appropriate lenses (interpretations). The impressionists used such a lens to understand the world (cf. the perspective of Vargas Llosa later in this chapter). Similarly, at first interpretations of some scientists were questioned and later the scientific community accepted them (for details, see Chapter 6). Science curriculum (and textbooks) can provide examples from the history of science to engage students creatively and thus form part of a responsible citizenry. In order to engage students creatively with science, and dream about discoveries and inventions, Perl (2007) has called for 'freedom to try out new ideas' and this requires less course and lab work.

f. These examples of the views of scientists, Nobel laureates, science educators, and in–servicescience teachers (after having been exposed to the HPS and the NOS), constitute almost a blueprint for a new curriculum that could motivate students to understand science and thus form part of a responsible citizenry: textbooks can be written differently → from rhetoric to understanding → experimental details accompanied by historical reconstructions → contradictions between scientists → motivation for analyzing and understanding other experiments → freedom to try out new ideas → dream about discoveries and inventions. It is concluded that the role of 'science in the making' in science instruction would be more effective if teachers deliberately scaffold classroom experiences based on a critical appraisal of an HPS framework.

Final Thoughts

In his Nobel Prize for Literature acceptance speech, Mario Vargas Llosa declared on December 10, 2010:

> La literatura es una representación falaz de la vida que, sin embargo, nos ayuda a entenderla mejor, a orientarnos por el laberinto en el que nacimos, transcurrimos y a morimos. *(Vargas Llosa, 2010; Literature is a false representation of life that nevertheless helps us to understand life better, to orient ourselves in the labyrinth where we are born, pass by, and die.)*

Let us now compare this with Leon Cooper's (Nobel laureate in physics) analogy between a style of painting (impressionism) and progress in science:

> I believe, in some ways, the scientist can be compared to the painter. The impressionists, for example, were accused of *not being able to see things as they are. (Reproduced in Niaz et al., 2010b, p. 48, italics added)*

In my opinion, there is an underlying thread that unites the thinking of Vargas Llosa and Cooper. If the impressionists were accused of *not being able to see things as they are*, it means that they had a false representation of the reality. However, as time goes by most people now see the world as the impressionists had conceptualized and this helps to provide a better vision of the world. Similarly, Vargas Llosa considers literature to provide a *false representation of life*. However, once again as time goes by this "false" representation helps us to understand life better. There is a striking and wonderful resemblance in the thinking of the two Nobel laureates, one a theoretical physicist and the other a novelist—perhaps the two could be considered as poets, looking for the essence of life in different endeavors of human pursuit.

REFERENCES

AAAS, American Association for Advancement of Science (1993). *Benchmarks for scientific literacy*. Washington, DC: AAAS.

AAPT, American Association of Physics Teachers (1996). *Powerful ideas in the physical sciences*. Washington, DC: Author.

Abd-El-Khalick,F. (2005). Developing deeper understandings of nature of science: The impact of a philosophy of science course on preservice science teachers' views and instructional planning. *International Journal of Science Education, 27*, 15–42.

Abd-El-Khalick,F., & Akerson, V.L. (2004). Learning about nature of science as conceptual change: Factors that mediate the development of preservice elementary teachers' views of science. *Science Education, 88*, 785–810.

Abd-El-Khalick,F., & Akerson, V.L. (2007). On the role and use of "theory" in science education research: A response to Johnston, Akerson and Sowell. *Science Education, 91*, 187–194.

Abd-El-Khalick,F., Bell, R.L., & Lederman, N.G. (1998). The nature of science and instructional practice: Making the unnatural natural. *Science Education, 82*, 417–436.

Abd-El-Khalick,F., BouJaoude, S., Duschl, R., Lederman, N.G., Mamlok-Naaman,R., Hofstein, A., Niaz, M., Treagust, D., & Tuan, H.-L.(2004). Inquiry in science education: International perspectives. *Science Education, 88*, 397–419.

Abd-El-Khalick,F., & Lederman, N.G. (2000). The influence of history of science courses on students' views of nature of science. *Journal of Research in Science Teaching, 37*, 1057–1095.

Abd-El-Khalick,F., Waters, M., & Le, A. (2008). Representation of nature of science in high school chemistry textbooks over the past four decades. *Journal of Research in Science Teaching, 45*, 835–855.

Abrams, G.S., & Perl, M.L. (1979). Measurement of the branching fraction for tau → rho nu. *Physics Review Letters, 43*, 1555–1558.

Achinstein, P. (1991). *Particles and waves: Historical essays in the philosophy of science*. New York: Oxford University Press.

Adey, P., & Shayer, M. (1994). *Really raising standards: Cognitive intervention and academic achievement*. London: Routledge.

Aikenhead, G., & Ryan, A.G. (1992). The development of a new instrument: "Views on science–technology–society"(VOSTS). *Science Education, 76*, 477–491.

Akerson, V.L., & Abd-El-Khalick, F. (2003). Teaching elements of nature of science: A year long case study of a fourth grade teacher. *Journal of Research in Science Teaching, 40*, 1025–1049.

Akerson, V.L., Abd-El-Khalick, F., & Lederman, N.G. (2000). Influence of a reflective explicit activity-based approach on elementary teachers' conceptions of nature of science. *Journal of Research in Science Teaching, 37*, 295–317.

Akerson, V.L., Buzzelli, C.A., & Donnelly, L.A. (2008). Early childhood teachers' views of nature of science: The influence of intellectual levels, cultural values, and explicit reflective teaching. *Journal of Research in Science Teaching, 45*, 748–770.

Akerson, V.L., & Hanuscin, D.L. (2007). Teaching nature of science through inquiry: Results of a 3-year professional development program. *Journal of Research in Science Teaching, 44*, 653–680.

Akerson, V.L., Morrison, J.A., & Roth McDuffie, A. (2006). One course is not enough: Preservice elementary teachers' retention of improved views of nature of science. *Journal of Research in Science Teaching, 43*, 194–213.

Akerson, V.L., & Volrich, M.L. (2006). Teaching nature of science explicitly in a first-grade internship setting. *Journal of Research in Science Teaching, 43*, 377–394.

Allchin, D. (2003). Scientific myth conceptions. *Science Education, 87*, 329–351.

Allchin, D. (2004a). The transparency of myth-making? *Science Education, 88*, 759–761.

Allchin, D. (2004b). Should the sociology of science be rated X? *Science Education, 88*, 934–946.

Alters, B.J. (1997). Whose nature of science? *Journal of Research in Science Teaching, 34*, 39–55.

Alvarez, W., & Azaro, F. (1990). An extraterrestrial impact. *Scientific American, 263*, 78–84.

Andrews, T. (1869). On the continuity of the gaseous and liquid states of matter. (First Bakerian Lecture, 1869. Reproduced in *Scientific papers*, pp. 296–317, London, 1896).

Arriassecq, I., & Greca, I.M. (2007). Approaches to the teaching of special relativity theory in high school and university textbooks of Argentina. *Science & Education, 16*, 65–86.

Bartholomew, H., Osborne, J., & Ratcliffe, M. (2004). Teaching students "ideas-about-science": Five dimensions of effective practice. *Science Education, 88*, 655–682.

Bauer, H.H. (1994). Scientific literacy and the myth of the scientific method. Champaign: University of Illinois Press.

Bell, R., Abd-El-Khalick, F., Lederman, N.G., McComas, W.F., & Matthews, M.R. (2001). The nature of science and science education: A bibliography. *Science & Education, 10*, 187–204.

Bencze, L., & Elshof, L. (2004). Science teachers as metascientists: An inductive-deductive dialectic immersion in northern alpine field ecology. *International Journal of Science Education, 26*, 1507–1526.

Bencze, L., & Hodson, D. (1999). Changing practice by changing practice: Toward more authentic science and science curriculum development. *Journal of Research in Science Teaching, 36*, 521–539.

Bevilacqua, F., & Bordoni, S. (1998). New contents for new media: Pavia project physics. *Science & Education, 7*, 451–469.

Bianchini, J.A., & Colburn, A. (2000). Teaching the nature of science through inquiry to prospective elementary teachers: A tale of two researchers. *Journal of Research in Science Teaching, 37*, 177–209.

Biao, L., Cai, T., Koenig, K., Fang, K., Han, J., Wang, J., Liu, Q., Ding, L., Cui, L., Luo, Y., Wang, Y., Li, L., & Wu, N. (2009). Learning and scientific reasoning. *Science, 232*, 586–587.

Bickard, M.H. (1997). Constructivisms and relativisms: A shopper's guide. *Science & Education, 6*, 29–42.

Blanco, R., & Niaz, M. (1997). Epistemological beliefs of students and teachers about the nature of science: From "Baconian inductive ascent" to the "irrelevance" of scientific laws. *Instructional Science, 25*, 203–231.

Bodner, G., & Pardue, H. (1989). *Chemistry: An experimental science*. New York: Wiley.

Bohr, N. (1913). On the constitution of atoms and molecules, Part I. *Philosophical Magazine, 26* (Series 6), 1–25.

Boulter, C.J., & Gilbert, J.K. (1995). Argument and science education. In P.J.M. Costello & S. Mitchell (Eds.), *Competing and consensual voices: The theory and practice of argumentation* (pp. 84–98). Clevedon, UK: Multilingual Matters.

Bricker, L.A., & Bell, P. (2008). Conceptualizations of argumentation from science studies and the learning sciences and their implications for the practice of science education. *Science Education, 92*, 473–498.

Brickhouse, N.W., Dagher, Z.R., Letts, W.J., & Shipman, H.L. (2000). Diversity of students' views about evidence, theory, and the interface between science and religion in an astronomy course. *Journal of Research in Science Teaching, 37*, 340–362.

Brito, A., Rodríguez, M.A., & Niaz, M. (2005). A reconstruction of development of the periodic table based on history and philosophy of science and its implications for general chemistry textbooks. *Journal of Research in Science Teaching, 42*, 84–111.

Brock, W.H. (1993). *The Norton history of chemistry*. New York: Norton.

Brush, S.G. (1976). *The kind of motion we call heat: A history of the kinetic theory of gases in the 19th century*. New York: North-Holland.

Brush, S.G. (1978). Why chemistry needs history and how it can get some. *Journal of College Science Teaching, 7*, 288–291.

Brush, S.G. (1996). The reception of Mendeleev's periodic law in America and Britain. *Isis, 87*, 595–628.

Burbules, N.C., & Linn, M.C. (1991). Science education and philosophy of science: Congruence or contradiction? *International Journal of Science Education, 13*, 227–241.

Campanario, J.M. (1999). La ciencia que no enseñamos. *Enseñanza de las Ciencias, 17*, 397–410.

Campanario, J.M. (2002). The parallelism between scientists' and students' resistance to new scientific ideas. *International Journal of Science Education, 24*, 1095–1110.

Campanario, J.M. (2004). Algunas posibilidades del artículo de investigación como recurso didáctico orientado a cuestionar ideas inadecuadas sobre la ciencia. *Enseñanza de las Ciencias, 22*, 365–378.

Campanario, J.M., y Otero, J.C. (2000). Más allá de las ideas previas como dificultades de aprendizaje: Las pautas de pensamiento, las concepciones epistemológicas y las estrategias metacognitivas de los alumnos de ciencias. *Enseñanza de las Ciencias, 18*, 155–169.

Campbell, D.T. (1988a). The experimenting society. In E.S. Overman (Ed.), *Methodology and epistemology for social science* (pp. 290–314). Chicago: University of Chicago Press (first published 1971).

Campbell, D.T. (1988b). Can we be scientific in applied social science? In E.S. Overman

(Ed.), *Methodology and epistemology for social science* (pp. 315–333). Chicago: University of Chicago Press (first published 1984 in *Evaluation Studies Review Annual*).
Cartwright, N. (1983). *How the laws of physics lie*. Oxford: Clarendon Press.
Cartwright, N. (1989). *Nature's capacities and their measurement*. Oxford: Clarendon Press.
Cartwright, N. (1999). *The dappled world: A study of the boundaries of science*. Cambridge: Cambridge University Press.
Cavalli-Sforza, M., et al. (1976). Anomalous production of high-energy muons in e^+e^- collisions at 4.8 GeV. *Physics Review Letters, 36*, 558–561.
Chamizo, J.A., & Gutiérrez, M.Y. (2004). Conceptos fundamentales en química 1. Valencia. *Educación Química, 15*, 359–365.
Chang, H. (2011). How historical experiments can improve scientific knowledge and science education: The cases of boiling water and electrochemistry. *Science & Education, 20*(3–4), 317–341.
Chang, R. (1981). *Chemistry*. New York: Random House.
Chang, R. (2007). *Chemistry* (9th Spanish ed.). New York: McGraw-Hill.
Chang, S.-N.,& Chiu, M.-H.(2008). Lakatos' scientific research programmes as a framework for analyzing informal argumentation about socio-scientific issues. *International Journal of Science Education, 30*, 1753–1773.
Chang, Y.-H., Chang, C.-Y., & Tseng, Y.-H. (2010). Trends of science education research: An automatic content analysis. *Journal of Science Education and Technology, 19*(4), 315–331.
Chen, S. (2006). Development of an instrument to assess views on nature of science and attitudes toward teaching science. *Science Education, 90*, 803–819.
Chiappetta, E.L., Sethna, G.H., & Fillman, D.A. (1991). A quantitative analysis of high school chemistry textbooks for scientific literacy themes and expository learning aids. *Journal of Research in Science Teaching, 28*, 939–951.
Chinn, C.A., & Anderson, R.C. (1998). The structure of discussions that promote reasoning. *Teachers College Record, 100*, 315–368.
Chinn, C.A., & Brewer, W.F. (1993). The role of anomalous data in knowledge acquisition: A theoretical framework and implications for science instruction. *Review of Educational Research, 63*, 1–49.
Chinn, C.A., & Malhotra, B. (2002). Epistemologically authentic inquiry in schools: A theoretical framework for evaluating inquiry tasks. *Science Education, 86*, 175–218.
Christie, M. (1994). Philosophers versus chemists concerning 'laws of nature', *Studies in History and Philosophy of Science, 25*, 613–629.
Christie, J.R., & Christie, M. (2003). Chemical laws and theories: A response to Vihalemm. *Foundations of Chemistry, 5*, 165–174.
Clark, D.B., & Sampson, V.D. (2007). Personally-seeded discussions to scaffold online argumentation. *International Journal of Science Education, 29*, 253–277.
Clark, P. (1976). Atomism versus thermodynamics. In C. Howson (Ed.), *Method and appraisal in the physical sciences: The critical background to modern science, 1800–1905* (pp. 41–105). Cambridge, UK: Cambridge University Press.
Clough, M.P. (2006). Learners' responses to the demands of conceptual change: Considerations for effective nature of science instruction. *Science & Education, 15*, 463–494.
Clough, M.P., & Olson, J.K. (2008). Teaching and assessing the nature of science: An introduction. *Science & Education, 17*, 143–145.
Cobern, W.W. (1992). *World view theory and science education research* (NARST monograph No. 3). Manhattan, KS: National Association for Research in Science Teaching.

Cobern, W.W., Gibson, A.T., & Underwood, S.A. (1999). Conceptualizations of nature: An interpretive study of 16 ninth graders' everyday thinking. *Journal of Research in Science Teaching, 36*, 541–564.

Cobern, W.W., & Loving, C.C. (2008). An essay for educators: Epistemological realism really is common sense. *Science & Education, 17*, 425–447.

Coll, R.K., & Taylor, N. (2004). Probing scientists' beliefs: How open–minded are modern scientists? *International Journal of Science Education, 26*, 757–778.

Collins, H. (2000). On beyond 2000. *Studies in Science Education, 35*, 169–173.

Collins, H. (2007). The uses of sociology of science for scientists and educators. *Science & Education, 16*, 217–230.

Collins, H., & Pinch, T. (1993). *The Golem: What everyone should know about science*. New York: Cambridge University Press.

Cooper, L.N. (1970). *An introduction to the meaning and structure of physics* (short edition). New York: Harper & Row.

Cooper, L.N. (1992). *Physics: Structure and meaning*. Hanover, NH: University Press of New England.

Cowan, C.L., et al. (1956). Detection of the free neutrino: A confirmation. *Science, 124*, 103–104.

Crawford, B.A., Zembal–Saul,C., Munford, D., & Friedrichsen, P. (2005). Confronting prospective teachers' ideas of evolution and scientific inquiry using technology and inquiry–basedtasks. *Journal of Research in Science Teaching, 42*, 613–637.

Crookes, W. (1877). The chemistry of the future. *Quarterly Journal of Science*, N.S., 7, 289–306.

Crowther, J.G. (1910). *Proceedings of the Royal Society* (vol. lxxxiv). London: Royal Society.

Cushing, J.T. (1978). Post–usereview of Leon N. Cooper's *An introduction to the meaning and structure of physics*, New York: Harper & Row. *American Journal of Physics, 46*(1), 114–116.

Cushing, J.T. (1989). The justification and selection of scientific theories. *Synthese, 78*, 1–24.

Dagher, Z.R.., & BouJaoude, S. (1997). Scientific views and religious beliefs of college students: The case of biological evolution. *Journal of Research in Science Teaching, 34*, 429–445.

Dagher, Z.R., & BouJaoude, S. (2005). Students' perceptions of the nature of evolutionary theory. *Science Education, 89*, 378–391.

Dagher, Z.R., Brickhouse, N.W., Shipman, H., & Letts, W.J. (2004). How some college students represent their understandings of the nature of scientific theories. *International Journal of Science Education, 26*, 735–755.

Darrigol, O. (1993). Strangeness and soundness in Louis de Broglie's early works. *Physis: Rivista Internazionale de Storia della Scienza, 30*(2–3), 303–372.

Davidsson, E., & Jakobsson, A. (2007). Different images of science at Nordic science centres. *International Journal of Science Education, 29*, 1229–1244.

Davisson, C., & Germer, L.H. (1927). Diffraction of electrons by a crystal of Nickel. *Physical Review, 30*(6), 705–740.

Dawid, R. (2006). Underdetermination and theory succession from the perspective of string theory. *Philosophy of Science, 73*, 298–322.

de Berg, K.C. (1989). The emergence of quantification in the pressure–volumerelation–ship for gases: A textbook analysis. *Science Education, 73*(2), 115–134.

de Berg, K.C. (2006). The kinetic–molecularand thermodynamic approaches to osmotic pressure: A study of dispute in physical chemistry and the implications for chemistry education. *Science & Education, 15*, 495–519.

de Berg, K.C. (2008a). The concepts of heat and temperature: The problem of determining the content for the construction of an historical case study which is sensitive to nature of science issues and teaching–learningissues. *Science & Education, 17*, 75–114.

de Berg, K.C. (2008b). Tin oxide chemistry from Macquer (1758) to Mendeleeff (1891) as revealed in the textbooks and other literature of the era. *Science & Education, 17*, 265–287.

de Berg, K.C. (2010). Tin oxide chemistry from the last decade of the nineteenth century to the first decade of the twenty–firstcentury: Towards the development of a big–picture approach to the teaching and learning of chemistry while focusing on a specific compound or class of compounds. *Science & Education, 19*(9), 847–866.

de Broglie, L. (1922). *Journal de Physique, 3* (Series VI), 422.

de Broglie, L. (1923a). Ondes et quanta. *Comptes Rendus, 177*, 507–510, 548–550, 630–632.

de Broglie, L. (1923b). Waves and quanta. *Nature, 112*, 540.

de Broglie, L. (1924). A tentative theory of light quanta. *Philosophical Magazine, 47*(Series 6), 446–458.

Delamont, S. (1990). A paradigm shift in research on science education? *Studies in Science Education, 18*, 153–177.

Delval, J. (2002). Entrevista a Juan Delval, realizada por P. Cañal. *Investigación en la Escuela, 43*, 71–80.

De Milt, C. (1951). The congress at Karlsruhe. *Journal of Chemical Education, 28*, 421–425.

Demircioğlu, H., Demircioğlu, G., & Çalik, A. (2009). Investigating the effectiveness of storylines embedded within a context–basedapproach: The case for the periodic table. *Chemistry Education Research and Practice, 10*, 241–249.

Deniz, H., Donnelly, L.A., & Yilmaz, I. (2008). Exploring the factors related to acceptance of evolutionary theory among preservice biology teachers: Toward a more informative conceptual ecology for biological evolution. *Journal of Research in Science Teaching, 45*, 420–443.

de Posada, J.M. (1999). Concepciones de los alumnos sobre el enlace químico antes, durante y después de la enseñanza formal. Problemas de aprendizaje. *Enseñanza de las Ciencias, 17*, 227–245.

Dickerson, R.E., Gray, H.B., Darensbourg, M.Y., & Darensbourg, D.J. (1984). *Chemical principles* (4th ed.). Menlo Park, CA: Benjamin Cummings.

Dmitriev, I.S. (2004). Scientific discovery *in statu nascendi*: The case of Dmitrii Mendeleev's periodic law. *Historical Studies in the Physical Sciences, 34*, 233–275.

Dogan, N., & Abd–El–KhalickF. (2008). Turkish grade 10 students' and science teachers' conceptions of nature of science: A national study. *Journal of Research in Science Teaching, 45*(10), 1083–1112.

Driver, R., Leach, J., Millar, R., & Scott, P. (1996). *Young people's images of science*. Milton Keynes, UK: Open University Press.

Driver, R., Newton, P., & Osborne, J. (2000). Establishing the norms of scientific argumentation in classrooms. *Science Education, 84*, 287–312.

Duhem, P. (1914). *The aim and structure of physical theory* (2nd ed., trans. Philip P. Wiener). New York: Atheneum.

Duschl, R.A. (1990). *Restructuring science education: The importance of theories and their development.* New York: Teachers College Press.

Duschl, R.A., & Osborne, J. (2002). Supporting and promoting argumentation discourse. *Studies in Science Education, 38*, 39–72.

Egan, K. (1983). *Education and psychology. Plato, Piaget and scientific psychology.* New York: Teachers College Press.

Ehrenhaft, F. (1941). The microcoulomb experiment. *Philosophy of Science, 8*, 403–457.

Einstein, A. (1905). Über einen Erzeugung und Verwandlung des Lichtes betreffenden heuristischen Gesichtspunkt. *Annalen der Physik, 17*, 132–148.

Einstein, A. (1909a). Zum gegenwärtigen stand des strahlungsproblems. *Physikalische Zeitschrift, 10*, 185–193.

Einstein, A. (1909b). Über die Entwicklung unsere anschauungen über das wesen und die konstitution der strahlung. *Physikalische Zeitschrift, 10*, 817–825.

Einstein, A. (1925). Quantentheorie des einatomigen idealen gases. 2. Abhandlung. *Preuss. Akad. d. Wiss., Sitzingsber*, 3–14.

Enfield, M., Smith, E.L., & Grueber, D.J. (2008). "A sketch is like a sentence": Curriculum structures that support teaching epistemic practices of science. *Science Education, 92*, 608–630.

Erduran, S. (2007). Breaking the law: Promoting domain-specificityin chemical education in the context of arguing about the periodic law. *Foundations of Chemistry, 9*, 247–263.

Erduran, S., & Jiménez-Aleixandre,M.P. (Eds.). (2008). *Argumentation in science education: Perspectives from classroom-based research.* Dordrecht, the Netherlands: Springer.

Erduran, S., Simon, S., & Osborne, J. (2004). TAPping into argumentation: Developments in the application of Toulmin's argument pattern for studying science discourse. *Science Education, 88*, 915–933.

Eshach, H. (2009). The Nobel prize in the physics class: Science, history and glamour. *Science & Education, 18*, 1377–1393.

Falconer, I. (1987). Corpuscles, electrons, and cathode rays: J. J. Thomson and the "discovery of the electron." *British Journal for the History of Science, 20*, 241–276.

Fara, P. (2009). *Science: A four thousand year history.* New York: Oxford University Press.

Feldman, G.J., & Perl, M.L. (1977). Inclusive anomalous muon production in e^+e^- annihilation. *Physics Review Letters, 38*, 117–120.

Fernández, I., Gil, D., Carrascosa, J., Cachapuz, A., & Praia, J. (2002). Visiones deformadas de la ciencia transmitidas por la enseñanza. *Enseñanza de las Ciencias, 20*, 477–488.

Feyerabend, P. (1993). *Against method.* New York: Verso.

Fine, L., & Beall, H. (1990). *Chemistry for engineers and scientists.* Philadelphia: Saunders College Publishing.

Florence, M.K., & Yore, L.D. (2004). Learning to write like a scientist: Co-authoringas an enculturation task. *Journal of Research in Science Teaching, 41*, 637–668.

Ford, M. (2008). Disciplinary authority and accountability in scientific practice and learning. *Science Education, 92*, 404–423.

Ford, M., & Wargo, B.M. (2007). Routines, roles, and responsibilities for aligning scientific and classroom practices. *Science Education, 91*, 133–157.

Frické, M. (1976). The rejection of Avogadro's hypothesis. In C. Howson (Ed.), *Method and appraisal in the physical sciences: The critical background to modern science, 1800–1905* (pp. 277–307). Cambridge, UK: Cambridge University Press.

Fujii, K. (1986). The Berthollet–Proustcontroversy and Dalton's chemical atomic theory 1800–1820. *British Journal for the History of Science, 19*, 177–200.

Furió, C., Azcona, R., & Guisasola, J. (2002). Revisión de investigaciones sobre la enseñanza–aprendizaje de los conceptos *cantidad de sustancia y mol. Enseñanza de las Ciencias, 20*, 229–242.

Gage, N. (1989). The paradigm wars and their aftermath. *Educational Researcher, 18*, 4–10.

Galagovsky, L. (Ed.). (2011). *Didáctica de las ciencias naturales: El caso de los modelos científicos.* Buenos Aires: Lugar Editorial.

García, J.J. (2000). La solución de situaciones problemáticas: Una estrategia didáctica para la enseñanza de la química. *Enseñanza de las Ciencias, 18*, 113–129.

Gavroglu, K. (2000). Controversies and the becoming of physical chemistry. In P. Machamer, M. Pera, & A. Baltas (Eds.), *Scientific controversies: Philosophical and historical perspectives* (pp. 177–198). New York: Oxford University Press.

Gay–Lussac, J.–L. (1808). Memoir on the combination of gaseous substances with each other. *Alembic Club Reprints, 4* (Edinburgh, 1923).

Gee, J. (1999). *An introduction to discourse analysis.* New York: Routledge.

Geiger, H., & Marsden, E. (1909). On a diffuse reflection of the alpha particles. *Proceedings of the Royal Society* (vol. lxxxii). London: Royal Society.

Gell–Mann, M. (1971). How scientists can really help. *Journal of College Science Teaching, 24*, 23–25.

Giere, R.N. (1999). *Science without laws.* Chicago: University of Chicago Press.

Giere, R.N. (2006). *Scientific perspectivism.* Chicago: University of Chicago Press.

Gillespie, R.J., Humphreys, D.A., Baird, N.C., & Robinson, E.A. (1989). *Chemistry.* Boston: Allyn and Bacon.

Gil–Pérez, D., Guisáosla, J., Moreno, A., Cachapuz, A., Pessoa de Carvalho, A.M., Martínez Torregrosa, J., et al. (2002). Defending constructivism in science education. *Science & Education, 11*, 557–571.

Gooday, G., Lynch, J.M., Wilson, K.G., & Barsky, C.K. (2008). Does science education need the history of science? *Isis, 99*, 322–330.

Gordin, M.D. (2002). The organic roots of Mendeleev's periodic law. *Historical Studies in the Physical Sciences, 32*, 263–290.

Gordin, M.D. (2004). *A well ordered thing: Dmitrii Mendeleev and the shadow of the periodic table.* New York: Basic Books.

Grandy, R., & Duschl, R.A. (2007). Reconsidering the character and role of inquiry in school science: Analysis of a conference. *Science & Education, 16*, 141–166.

Grosser, M. (1962). *The discovery of Neptune.* Cambridge, MA: Harvard University Press (Reprint, New York: Dover, 1979).

Guba, E.G., & Lincoln, Y.S. (1989). Fourth generation evaluation. Newbury Park, CA: Sage.

Guba, E.G., & Lincoln, Y.S. (2005). Paradigmatic controversies, contradictions, and emerging confluences. In N.K. Denzin & Y.S. Lincoln (Eds.), *The Sage handbook of qualitative research* (3rd ed., pp. 191–215). Thousand Oaks, CA: Sage.

Halliday, D., & Resnick, R. (1981). *Fundamentals of physics* (2nd ed.). New York: Wiley.

Hanson, N.R. (1958). *Patterns of discovery: An inquiry into the conceptual foundations of science.* Cambridge, UK: Cambridge University Press.

Hanson, N.R. (1964). On the structure of physical knowledge. In S. Elam (Ed.), *Education and the structure of knowledge* (pp. 148–187). Chicago: Rand McNally & Company.

Hanuscin, D.L., Akerson, V.L., Phillipson-Mower, T. (2006). Integrating nature of science instruction into physical science content course for preservice elementary teachers: NOS views of teaching assistants. *Science Education, 90*, 912–935.

Havdala, R., & Ashkenazi, G. (2007). Coordination of theory and evidence: Effect of epistemological theories on students' laboratory practice. *Journal of Research in Science Teaching, 44*, 1134–1159.

Hecht, E. (1998). *Physics: Algebra/trig* (2nd ed.). Pacific Grove, CA: Brooks Cole.

Heering, P., & Klassen, S. (2010). Doing it differently: Attempts to improve Millikan's oil-dropexperiment. *Physics Education, 45*(4), 382–393.

Heilbron, J.L. (1981). *Historical studies in the theory of atomic structure*. New York: Arno Press.

Heilbron, J.L., & Kuhn, T. (1969). The genesis of the Bohr atom. *Historical Studies in the Physical Sciences, 1*, 211–290.

Hettema, H. (1995). Bohr's theory of the atom 1913–1923: A case study in the progress of scientific research programmes. *Studies in History and Philosophy of Modern Physics, 26*, 307–323.

Hipkins, R., Barker, M., & Bolstad, R. (2005). Teaching the "nature of science": Modest adaptations or radical reconceptions? *International Journal of Science Education, 27*, 243–254.

Hodson, D. (1985). Philosophy of science, science, and science teaching. *Studies in Science Education, 12*, 25–57.

Hodson, D. (1988). Toward a philosophically more valid science curriculum. *Science Education, 72*, 19–40.

Hodson, D. (2008). *Towards scientific literacy: A teacher's guide to the history, philosophy and sociology of science*. Rotterdam: Sense Publishers.

Hodson, D. (2009). *Teaching and learning about science: Language, theories, methods, history, traditions and values*. Rotterdam: Sense Publishers.

Hogan, K., & Maglienti, M. (2001). Comparing the epistemological underpinnings of students' and scientists' reasoning about conclusions. *Journal of Research in Science Teaching, 38*, 663–687.

Hokayem, H., & BouJaoude, S. (2008). College students' perceptions of the theory of evolution. *Journal of Research in Science Teaching, 45*, 395–419.

Holbrook, J., & Rannikmae, M. (2007). The nature of science education for enhancing scientific literacy. *International Journal of Science Education, 29*, 1347–1362.

Holton, G. (1952). *Introduction to concepts and theories in physical science*. New York: Addison-Wesley.

Holton, G. (1969). Einstein and the "crucial" experiment. *American Journal of Physics, 37*, 968–982.

Holton, G. (1978). Subelectrons, presuppositions, and the Millikan-Ehrenhaft dispute. *Historical Studies in the Physical Sciences, 9*, 161–224.

Holton, G. (1988). On the hesitant rise of quantum physics research in the United States. In S. Goldberg & R.H. Stuewer (Eds.), *The Michelson era in American science, 1870–1930* (pp. 177–205). New York: American Institute of Physics.

Holton, G. (1992). Ernst Mach and the fortunes of positivism in America. *Isis, 83*, 27–60.

Holton, G. (1993). *Science and anti-science*. Cambridge, MA: Harvard University Press.

Holton, G. (1998). *The scientific imagination*. Cambridge, MA: Harvard University Press.

Holton, G. (1999). R.A. Millikan's struggle with the meaning of Planck's constant. *Physics in Perspective, 1*, 231–237.

Holton, G. (2003). The project physics course, then and now. *Science & Education, 12*, 779–786.

Holton, G., & Brush, S.G. (2001). *Physics, the human adventure: From Copernicus to Einstein and beyond* (3rd ed.). New Brunswick, NJ: Rutgers University Press.

Hosson, C., & Kaminski, W. (2007). Historical controversy as an educational tool: Evaluating elements of a teaching–learning sequence conducted with the text "Dialogues on the ways that vision operates." *International Journal of Science Education, 29*, 617–642.

Höttecke, D., Henke, A., & Riess, F. (2011). Implementing history and philosophy in science teaching: Strategies, methods, results and experiences from the European HIPST Project. *Science & Education, 20* (in press).

Hovis, R.C., & Kragh, H. (1991). Resource letter HEPP–1:History of elementary particle physics. *American Journal of Physics, 59*, 779–807.

Howe, K.R. (1988). Against the quantitative–qualitativeincompatibility thesis or dogmas die hard. *Educational Researcher, 17*, 10–16.

Ihde, A.J. (1964). *The development of modern chemistry*. New York: Harper & Row.

Inhelder, B., & Piaget, J. (1958). *The growth of logical thinking from childhood to adolescence*. New York: Basic Books (first published 1955).

Irez, S. (2006). Are we prepared? An assessment of preservice science teacher educators' beliefs about nature of science. *Science Education, 90*, 1113–1143.

Isaacson, W. (2007). *Einstein*. New York: Simon & Schuster.

Jammer, M. (1996). *The conceptual development of quantum mechanics*. New York: McGraw–Hill.

Jenkins, E. (2007). School science: A questionable construct? *Journal of Curriculum Studies, 39*(3), 265–282.

Johnson, R.B., & Onwuegbuzie, A.J. (2004). Mixed methods research: A research paradigm whose time has come. *Educational Researcher, 33*, 14–26.

Johnston, A., Southerland, S.A., & Sowell, S. (2006). Dissatisfied with the fruitfulness of "learning ecologies." *Science Education, 90*, 907–911.

Jones, R.C. (1995). The Millikan oil–dropexperiment: Making it worthwhile. *American Journal of Physics, 63*, 970–977.

Justi, R., & Gilbert, J. (1999). A cause of ahistorical science teaching: Use of hybrid models. *Science Education, 83*, 163–177.

Kaji, M. (2003). Mendeleev's discovery of the periodic law: The origin and the reception. *Foundations of Chemistry, 5*, 189–214.

Kang, N.–H., & Wallace, C.S. (2005). Secondary science teachers' use of laboratory activities: Linking epistemological beliefs, goals, and practices. *Science Education, 89*, 140–165.

Kang, S., Scharmann, L.C., & Noh, T. (2005). Examining students' views on the nature of science: Results from Korean 6th, 8th, and 10th graders. *Science Education, 89*, 314–334.

Kapusta, J.I. (1975). Best measuring time for a Millikan oil drop experiment. *American Journal of Physics, 43*, 799–800.

Karmiloff–Smith,A., & Inhelder, B. (1975). "If you want to get ahead, get a theory." *Cognition, 3*, 195–212.

Kawasaki, K., Herrenkohl, L.R., & Yeary, S.A. (2004). Theory building and modeling in a sinking and floating unit: A case study of third and fourth grade students' developing epistemologies of science. *International Journal of Science Education, 26*, 1299–1324.

Kelly, G.J., Drucker, S., & Chen, C. (1998). Students' reasoning about electricity: Combining performance assessments with argumentation analysis. *International Journal of Science Education, 20*, 849–871.

Kesidou, S., & Roseman, J.E. (2002). How well do middle school science programs measure up? Findings from Project 2061's curricular review. *Journal of Research in Science Teaching, 39*, 522–549.

Khishfe, R. (2008). The development of seventh graders' views of nature of science. *Journal of Research in Science Teaching, 45*, 470–496.

Khishfe, R., & Lederman, N.G. (2006). Teaching nature of science within a controversial topic: Integrated versus non-integrated.*Journal of Research in Science Teaching, 43*, 395–418.

Khishfe, R., & Lederman, N. (2007). Relationship between instructional context and views of nature of science. *International Journal of Science Education, 29*, 939–961.

Kittleson, J.M., & Southerland, S.A. (2004). The role of discourse in group knowledge construction: A case study of engineering students.*Journal of Research in Science Teaching, 41*, 267–293.

Klassen, S. (2006). A theoretical framework for contextual science teaching. *Interchange, 37*, 31–62.

Klassen, S. (2009). Identifying and addressing student difficulties with the Millikan oil drop experiment. *Science & Education, 18*, 593–607.

Klassen, S. (2011). The photoelectric effect: Reconstructing the story for the physics classroom. *Science & Education, 20*(7–8), 719–731.

Klassen, S., & Dietrich, S. (2010). Physics comes to Winnipeg: The 1909 meeting of the British Association for the Advancement of Science. *Interchange, 41*, 407–423.

Klein, M.J. (1964). Einstein and wave-particle duality. *The Natural Philosopher, 3*, 3–49.

Klopfer, L.E. (1969). The teaching of science and the history of science. *Journal of Research in Science Teaching, 6*, 87–95.

Koliopoulos, D., & Constantinou, C. (2005). The pendulum as presented in school science textbooks of Greece and Cyprus. *Science & Education, 14*, 59–73.

Kolstø, S.D. (2008). Science education for democratic citizenship through the use of history of science. *Science & Education, 17*, 977–997.

Koslowski, B. (1996). Theory and evidence: The development of scientific reasoning. Cambridge, MA: MIT Press.

Kotz, J.C., & Purcell, K.F. (1991). *Chemistry and chemical reactivity* (2nd ed.). Philadelphia: Saunders.

Kousathana, M., Demerouti, M., & Tsaparlis, G. (2005). Instructional misconceptions in acid-baseequilibria: An analysis from a history and philosophy of science perspective. *Science & Education, 14*, 173–193.

Kragh, H. (1999). *Quantum generations: A history of physics in the twentieth century*. Princeton, NJ: Princeton University Press.

Krane, K.S. (1996). *Modern physics* (2nd ed.). New York: Wiley.

Kruglak, H. (1972). Another look at the Pasco-Millikan oil-drop apparatus. *American Journal of Physics, 40*, 768–769.

Kubli, F. (1970). *Archives of the History of the Exact Sciences, 7*, 26–68.

Kuhn, D. (1993). *The skills of argument*. Cambridge, UK: Cambridge University Press.

Kuhn, D. (1997). Developmental psychology and science education. *Review of Educational Research, 67*, 141–150.

Kuhn, T.S. (1962). *The structure of scientific revolutions*. Chicago: University of Chicago Press.

Kuhn, T.S. (1970). *The structure of scientific revolutions* (2nd ed.). Chicago: University of Chicago Press.

Kuhn, T.S. (1984). Revisiting Planck. *Historical Studies in the Physical Sciences, 14*, 231–252.

Kuhn, T.S. (1985). *The Copernican revolution*. Cambridge, MA: Harvard University Press (first published 1957).

Laburú, C.E., & Niaz, M. (2002). A Lakatosian framework to analyze situations of cognitive conflict and controversy in students' understanding of heat energy and temperature. *Journal of Science Education and Technology, 11*, 211–219.

Lakatos, I. (1970). Falsification and the methodology of scientific research programmes. In I. Lakatos & A. Musgrave (Eds.), *Criticism and the growth of knowledge* (pp. 91–195). Cambridge, UK: Cambridge University Press.

Laudan, L. (1977). *Progress and its problems: Towards a theory of scientific growth*. Berkeley, CA: University of California Press.

Laudan, R., Laudan, L., & Donovan, A. (1988). Testing theories of scientific change. In A. Donovan, L. Laudan, & R. Laudan (Eds.), *Scrutinizing science: Empirical studies of scientific change* (pp. 3–44). Dordrecht, the Netherlands: Kluwer.

Lawson, A.E. (2003). The nature and development of hypothetico-predictiveargumentation with implications for science teaching. *International Journal of Science Education, 25*, 1387–1408.

Lawson, A.E. (2005). What is the role of induction and deduction in reasoning and scientific inquiry? *Journal of Research in Science Teaching, 42*, 716–740.

Lawson, A.E. (2010). How "scientific" is science education research? *Journal of Research in Science Teaching, 47*(3), 257–275.

Leach, J., Driver, R., Millar, R., & Scott, P. (1977). A study of progression in learning about "the nature of science": Issues of conceptualization and methodology. *International Journal of Science Education, 19*, 147–166.

Lederman, N.G. (1992). Students' and teachers' conceptions of the nature of science: A review of the research. *Journal of Research in Science Teaching, 29*, 331–359.

Lederman, N.G., Abd-El-KhalickF., Bell, R.L., & Schwartz, R. (2002). Views of nature of science questionnaire: Toward valid and meaningful assessment of learners' conceptions of nature of science. *Journal of Research in Science Teaching, 39*, 497–521.

Lederman, N.G., Wade, P., & Bell, R.L. (1998). Assessing understanding of the nature of science: A historical perspective. In W.F. McComas (Ed.), *The nature of science in science education: Rationales and strategies* (pp. 331–350). Dordrecht, the Netherlands: Kluwer.

Lee, G., Kwon, J., Park, S., Kim, J., Kwon, H., & Park, H. (2003). Development of an instrument for measuring cognitive conflict in secondary-levelscience classes. *Journal of Research in Science Teaching, 40*, 585–603.

Lee, M.-H.,Wu, Y.-T.,Tsai, C.-C.(2009). Research trends in science education from 2003 to 2007: A content analysis of publications in selected journals. *International Journal of Science Education, 31*, 1999–2020.

Leite, L. (2002). History of science in science education: Development and validation of a checklist for analyzing the historical content of science textbooks. *Science & Education, 11*, 333–359.

Levere, T.H. (2006). What history can teach us about science: Theory and experiment, data and evidence. *Interchange, 37*, 115–128.

Lin, C.-C., & Tsai, C.-C. (2008). Exploring the structural relationships between high school students' scientific epistemological views and their utilization of information commitments toward online science information. *International Journal of Science Education, 30*(15), 2001–2022.

Lin, H., & Chen, C. (2002). Promoting preservice chemistry teachers' understanding about the nature of science through history. *Journal of Research in Science Teaching, 39*, 773–792.

Lin, H.-S.,& Chiu, H.-L.(2004). Research report. *International Journal of Science Education, 26*, 101–112.

Linn, M.C., Songer, N.B., & Lewis, E.L. (1991). Overview: Students' models and epistemologies of science. *Journal of Research in Science Teaching, 28*, 729–732.

Lipton, P. (1991). *Inference to the best explanation.* London: Routledge.

Lipton, P. (2005). Testing hypotheses: Prediction and prejudice. *Science, 307*(14 January), 219–221.

Liu, S.-Y.,& Lederman, N.G. (2007). Exploring prospective teachers' worldviews and conceptions of nature of science. *International Journal of Science Education, 29*, 1281–1307.

Liu, S.-Y.,& Tsai, C.-C. (2008). Differences in the scientific epistemological views of undergraduate students. *International Journal of Science Education, 30*(8), 1055–1073.

Longino, H.E. (1990). *Science as social knowledge.* Princeton, NJ: Princeton University Press.

Loving, C.C., & Cobern, W.A. (2000). Invoking Thomas Kuhn: What citation analysis reveals for science education. *Science & Education, 9*, 187–206.

Machamer, P., Darden, D., & Craver, C.F. (2000). Thinking about mechanisms. *Philosophy of Science, 67*, 1–25.

Machamer, P., Pera, M., & Baltas, A. (2000). Scientific controversies: An introduction. In P. Machamer, M. Pera, & A. Baltas (Eds.), *Scientific controversies: Philosophical and historical perspectives* (pp. 3–17). New York: Oxford University Press.

Machamer, P., & Wolters, G. (2004). Introduction. In P. Machamer & G. Wolters (Eds.), *Science, values and objectivity* (pp. 1–13). Pittsburgh: University of Pittsburgh Press.

Mahan, B., & Myers, R.J. (1987). *University chemistry* (4th ed., Spanish). Menlo Park, CA: Benjamin Cummings.

Maher, P. (1988). Prediction accommodation and the logic of discovery. In A. Fine and J. Leplin (Eds.), *PSA 1988*, vol. 1. East Lansing: Philosophy of Science Association.

Marcano, C. (2010). Una reconstrucción histórica de la dualidad onda-partículabasada en la historia y filosofía de la ciencia y sus implicaciones para los textos universitarios de química general. Master of Science (chemistry education) dissertation. Universidad de Oriente, Cumaná, Venezuela.

Marín, N., Solano, I., & Jiménez, E. (1999). Tirando del hilo de la madeja constructivista. *Enseñanza de las Ciencias, 17*, 479–492.

Marquit, E. (1978). Philosophy of physics in general physics courses. *American Journal of Physics, 46*, 784–789.

Marshall, J.A., & Young, E.S. (2006). Preservice teachers' theory development in physical and simulated environments. *Journal of Research in Science Teaching, 43*, 907–937.

Martín-Díaz, M.J. (2006). Educational background, teaching experience, and teachers' views on the inclusion of nature of science in the science curriculum. *International Journal of Science Education, 28*, 1161–1180.

Martínez, A. (1999). Constructivismo radical, marco teórico de investigación y enseñanza de las ciencias. *Enseñanza de las Ciencias, 17*, 493–502.

Martins, I., Mortimer, E., Osborne, J., Tsatsarelis, C., Jimènez-Aleixandre,M.P. (2001). Rhetoric and science education. In H. Behrendt et al. (Eds.), *Research in science education—past, present and future* (pp. 189–198). Dordrecht, the Netherlands: Kluwer.

Masterton, W., Slowinski, E., & Stanitski, C. (1986). *Chemical principles* (4th Spanish ed.). Philadelphia: Saunders College Publishing.

Matthews, M.R. (1990). History, philosophy, and science teaching: A rapprochement. *Studies in Science Education, 18*, 25–51.

Matthews, M.R. (1994a). *Science teaching: The contribution of history and philosophy of science.* New York: Routledge.

Matthews, M.R. (1994b). Historia, filosofía y enseñanza de las ciencias: La aproximación actual. *Enseñanza de las Ciencias, 12*, 255–277.

Matthews, M.R. (1998). In defense of modest goals when teaching about the nature of science. *Journal of Research in Science Teaching, 35*, 161–174.

Matthews, M.R. (2004). Thomas Kuhn's impact on science education: What lessons can be learned? *Science Education, 88*, 90–118.

Maxwell, J.C. (1860). Illustrations of the dynamical theory of gases. *Philosophical Magazine, 19*, 19–32. (Reproduced in *Scientific Papers*, 1965, pp. 377–409, New York: Dover.)

McComas, W.F. (1996). Ten myths of science: Reexamining what we think we know about the nature of science. *School Science and Mathematics, 96*, 10–16.

McComas, W.F. (2008). Seeking historical examples to illustrate key aspects of the nature of science. *Science & Education, 17*, 249–263.

McComas, W.F., Almazroa, H. & Clough, M.P. (1998). The role and character of the nature of science in science education. *Science & Education, 7*, 511–532.

Medawar, P.B. (1967). *The art of the soluble.* London: Methuen.

Medicus, H.A. (1974). Fifty years of matter waves. *Physics Today, 27*, 38–45.

Méheut, M. (1997). Designing a learning sequence about a prequantitative kinetic model of gases: The parts played by questions and by computer-simulation. *International Journal of Science Education, 19*, 647–660.

Meichtry, Y.J. (1993). The impact of science curriculum on student views about the nature of science. *Journal of Research in Science Teaching, 30*, 429–443.

Mellado, V. (2003). Cambio didáctico del profesorado de ciencias experimentales y filosofía de la ciencia. *Enseñanza de las Ciencias, 21*, 343–358.

Mellado, V., Ruiz, C., Bermejo, M.L., & Jiménez, R. (2006). Contributions from the philosophy of science to the education of science teachers. *Science & Education, 15*, 419–445.

Mendeleev, D. (1869). Ueber die beziehungen der eigenschaften zu den atom gewichten der elemente. *Zeitschrift für Chemie, 12*, 405–406 (English trans. by C. Giunta).

Mendeleev, D. (1889). The periodic law of the chemical elements. *Journal of the Chemical Society, 55*, 634–656 (Faraday lecture, delivered on June 4, 1889).

Metz, D. (2007). We now interrupt the story: Mediating student learning using historical stories. Paper presented at the 9th International History, Philosophy and Science Teaching Conference, Calgary, Canada, June.

Metz, K. (1997). On the complex relation between cognitive developmental research and children's science curricula. *Review of Educational Research, 67*, 151–163.

Millar, R., & Osborne, J.F. (Eds.). (1998). *Beyond 2000: Science education for the future.* London: King's College London.

Millikan, R.A. (1913). On the elementary electrical charge and the Avogadro constant. *Physical Review, 2*, 109–143.
Millikan, R.A. (1914). A direct determination of "*h.*" *Physical Review, 4*, 73–75.
Millikan, R.A. (1915). New tests of Einstein's photo-electric equation. *Physical Review, 6*, 55.
Millikan, R.A. (1916a). "Einstein's photoelectric equation and contact electromotive force." *Physical Review,* 7, 18–32.
Millikan, R.A. (1916b). A direct photoelectric determination of Planck's "*h.*" *Physical Review,* 7, 355–388.
Millikan, R.A. (1917). *The electron.* Chicago: University of Chicago Press.
Millikan, R.A. (1950). *The autobiography of Robert A. Millikan.* New York: Prentice-Hall.
Millikan, R.A. (1965). The electron and the light-quant from the experimental point of view (Nobel prize acceptance speech, 1923). *Nobel Lectures: Physics* (pp. 54–66). Amsterdam: Elsevier.
Monk, M., & Osborne, J. (1997). Placing the history and philosophy of science on the curriculum: A model for the development of pedagogy. *Science Education, 81*, 405–424.
Moore, F.J., & Hall, W.T. (1939). *A history of chemistry.* New York: McGraw-Hill.
Mortimer, E.F. (1995). Conceptual change or conceptual profile change? *Science & Education, 4*, 267–285.
Mortimer, E.F., & Machado, A.H. (2000). Anomalies and conflicts in classroom discourse. *Science Education, 84*, 429–444.
Moseley, H.G.J. (1913). High frequency spectra of the elements. *Philosophical Magazine, 26*, 1025–1034.
Moseley, H.G.J. (1914). High frequency spectra of the elements, Part II. *Philosophical Magazine, 27*, 703–713.
Motterlini, M. (Ed.). (1999). *For and against method: Including Lakatos's lectures on scientific method and the Lakatos-Feyerabend correspondence.* Chicago: University of Chicago Press.
Murphy, N. (1989). Another look at novel facts. *Studies in History and Philosophy of Science, 20*, 385–388.
Nagel, E. (1961). *The structure of science.* New York: Harcourt, Brace & World.
Nash, L.K. (1956). The origin of Dalton's chemical atomic theory. *Isis, 47*, 101–116.
National Research Council, NRC (2002). *Scientific research in education.* Washington, DC: National Academy Press.
National Society for the Study of Education (1947). *Science education in American schools: Forty-sixth yearbook of the NSSE.* Chicago: University of Chicago Press.
Newton, P., Driver, R., & Osborne, J. (1999). The place of argumentation in the pedagogy of school science. *International Journal of Science Education, 21*, 553–576.
Niaz, M. (1994). Enhancing thinking skills: Domain specific/domain general strategies—A dilemma for science education. *Instructional Science, 22*, 413–422.
Niaz, M. (1995a). Cognitive conflict as a teaching strategy in solving chemistry problems: A dialectic-constructivist perspective. *Journal of Research in Science Teaching, 32*, 959–970.
Niaz, M. (1995b). Progressive transitions from algorithmic to conceptual understanding in student ability to solve chemistry problems: A Lakatosian interpretation. *Science Education, 79*, 19–36.
Niaz, M. (1997). How early can children understand some form of "scientific reasoning?" *Perceptual and Motor Skills, 85*, 1272–1274.
Niaz, M. (1998). From cathode rays to alpha particles to quantum of action: A rational

reconstruction of structure of the atom and its implications for chemistry textbooks. *Science Education, 82*, 527–552.

Niaz, M. (2000a). A rational reconstruction of the kinetic molecular theory of gases based on history and philosophy of science and its implications for chemistry textbooks. *Instructional Science, 28*, 23–50.

Niaz, M. (2000b). The oil drop experiment: A rational reconstruction of the Millikan–Ehrenhaft controversy and its implications for chemistry textbooks. *Journal of Research in Science Teaching, 37*, 480–508.

Niaz, M. (2000c). A framework to understand students' differentiation between heat energy and temperature and its educational implications. *Interchange, 31*, 1–20.

Niaz, M. (2001a). Understanding nature of science as progressive transitions in heuristic principles. *Science Education, 85*, 684–690.

Niaz, M. (2001b). How important are the laws of definite and multiple proportions in chemistry and teaching chemistry?—A history and philosophy of science perspective. *Science & Education, 10*, 243–266.

Niaz, M. (2001c). A rational reconstruction of the origin of the covalent bond and its implications for general chemistry textbooks. *International Journal of Science Education, 23*(6), 623–641.

Niaz, M. (2002). Facilitating conceptual change in students' understanding of electro-chemistry. *International Journal of Science Education, 24*, 425–439.

Niaz, M. (2005a). An appraisal of the controversial nature of the oil drop experiment: Is closure possible? *British Journal for the Philosophy of Science, 56*, 681–702.

Niaz, M. (2005b). Por qué los textos de química general no cambian y siguen una "retórica de conclusiones?" *Educación Química, 16*, 410–415.

Niaz, M. (2006). Can the study of thermochemistry facilitate students' differentiation between heat energy and temperature? *Journal of Science Education and Technology, 15*(3), 269–276.

Niaz, M. (2008a). *Teaching general chemistry: A history and philosophy of science perspective.* New York: Nova Science Publishers.

Niaz, M. (2008b). What "ideas–about–science"should be taught in school science? A chemistry teachers' perspective. *Instructional Science, 36*, 233–249.

Niaz, M. (2008c). Whither constructivism?—A chemistry teachers' perspective. *Teaching and Teacher Education, 24*, 400–416.

Niaz, M. (2008d). A rationale for mixed methods (integrative) research programmes in education. *Journal of Philosophy of Education, 42*(2), 287–305.

Niaz, M. (2008e). *Physical science textbooks: History and philosophy of science perspective.* New York: Nova Science Publishers.

Niaz, M. (2009a). *Critical appraisal of physical science as a human enterprise: Dynamics of scientific progress.* Dordrecht, the Netherlands: Springer.

Niaz, M. (2009b). Progressive transitions in chemistry teachers' understanding of nature of science based on historical controversies. *Science & Education, 18*, 43–65.

Niaz, M. (2010a). Are we teaching science as practiced by scientists? *American Journal of Physics, 78*, 5–6.

Niaz, M. (2010b). Science curriculum and teacher education: The role of presuppositions, contradictions, controversies and speculations vs Kuhn's "normal science." *Teaching and Teacher Education, 26*, 891–899.

Niaz, M. (2011). *Innovating science teacher education: A history and philosophy of science perspective.* New York: Routledge.

Niaz, M., Abd-El-Khalick,F., Benarroch, A., Cardellini, L., Laburú, C.E., Marín, N., et al. (2003). Constructivism: Defense or a continual critical appraisal—A response to Gil-Pérezet al. *Science and Education, 12*, 787–797.

Niaz, M., Aguilera, D., Maza, A., & Liendo, G. (2002). Arguments, contradictions, resistances and conceptual change in students' understanding of atomic structure. *Science Education, 86*, 505–525.

Niaz, M., & Cardellini, L. (2011). What can the Bohr-Sommerfeldmodel show students of chemistry in the 21st century? *Journal of Chemical Education, 88*, 240–243.

Niaz, M., & Coştu, B. (2009). Presentation of atomic structure in Turkish general chemistry textbooks. *Chemistry Education Research and Practice, 10*, 233–240.

Niaz, M., & Fernández, R. (2008). Understanding quantum numbers in general chemistry textbooks. *International Journal of Science Education, 30*, 869–901.

Niaz, M., Klassen, S., McMillan, B., & Metz, D. (2010a). Reconstruction of the history of the photoelectric effect and its implications for general physics textbooks. *Science Education, 94*, 903–931.

Niaz, M., Klassen, S., McMillan, B., & Metz, D. (2010b). Leon Cooper's perspective on teaching science: An interview study. *Science & Education, 19*, 39–54.

Niaz, M., & Maza, A. (2011). *Nature of science in general chemistry textbooks*. Dordrecht, the Netherlands: Springer Briefs in Education.

Niaz, M., & Rodríguez, M.A. (2001). Do we have to introduce history and philosophy of science or is it already "inside" chemistry? *Chemistry Education: Research and Practice in Europe, 2*, 159–164.

Niaz, M., & Rodríguez, M.A. (2005). The oil drop experiment: Do physical chemistry textbooks refer to its controversial nature? *Science and Education, 14*, 43–57.

Niaz, M., Rodríguez, M.A., & Brito, A. (2004). An appraisal of Mendeleev's contribution to the development of the periodic table. *Studies in History and Philosophy of Science, 35*, 271–282.

NRC, National Research Council (1996). *National science education standards*. Washington, DC: National Academy Press.

NRC, National Research Council (2002). *Science research in education*. Washington, DC: National Academy Press.

Nussbaum, M.E., Sinatra, G.M., & Poliquin, A. (2008). Role of epistemic beliefs and scientific argumentation in science learning. *International Journal of Science Education, 30*(15), 1977–1999.

Olenick, R.P., Apostol, T.M., & Goodstein, D.L. (1985). *Beyond the mechanical universe. From electricity to modern physics*. New York: Cambridge University Press.

O'Neill, D.K., & Polman, J.L. (2004). Why educate "little scientists?" Examining the potential of practice-basedscientific literacy. *Journal of Research in Science Teaching, 41*, 234–266.

Osborne, J.F. (2007). Science education for the twenty first century. *Eurasia Journal of Mathematics, Science and Technology Education, 3*, 173–184.

Osborne, J., Collins, S., Ratcliffe, M., Millar, R., & Duschl, R. (2003). What "ideas-about-science"should be taught in school science? A Delphi study of the expert community. *Journal of Research in Science Teaching, 40*, 692–720.

Osborne, J., Erduran, S., & Simon, S. (2004). Enhancing the quality of argumentation in school science. *Journal of Research in Science Teaching, 41*, 994–1020.

Ospina, J. (2010). Efecto fotoeléctrico: Una reconstrucción racional basada en la historia y filosofía de la ciencia y sus implicaciones para los textos de química general.

Master of Science (chemistry education) dissertation, Universidad de Oriente, Cumaná, Venezuela.

Oulton, C., Dillon, J., & Grace, M.M. (2004). Reconceptualizing the teaching of controversial issues. *International Journal of Science Education, 26*, 411–423.

Oxtoby, D., Nachtrieb, N., & Freeman, W. (1990). *Chemistry: Science of change* (2nd ed.). Philadelphia: Saunders College Publishing.

Padilla, K., & Furio-Mas,C. (2008). The importance of history and philosophy of science in correcting distorted views of "amount of substance" and "mole" concepts in chemistry teaching. *Science & Education, 17*, 403–424.

Paraskevopoulou, E., & Koliopoulos, D. (2011). Teaching nature of science through the Millikan-Ehrenhaftdispute. *Science & Education, 20*, in press.

Park, J. (2006). Modelling analysis of students' processes of generating scientific explanatory hypotheses. *International Journal of Science Education, 28*, 469–489.

Paschen, F. (1916). *Annalen der Physik*, 50, 901–940.

Patsopoulos, D. (2008). What Archimedes found with "eureka"? The interpretations of the history of physics in modern Greek textbooks of physics (18th–19th centuries). In M.V. Thomase (Ed.), *Science education in focus* (pp. 275–290). New York: Nova Science Publishers.

Pauling, L. (1952). *College chemistry*. San Francisco: Freeman.

Pauling, L. (1964). *General chemistry* (3rd ed.). San Francisco: Freeman.

Penrose, R. (1994). *Shadows of the mind: A search for the missing science of consciousness*. New York: Oxford University Press.

Perl, M.L. (1995). Martin L. Perl autobiography. (http://www.nobel.se/physics/laureates/1995/perl-autobio.html,downloaded October 10, 2002).

Perl, M.L. (1997). Reflections on the discovery of the Tau Lepton. In G. Ekspong (Ed.), *Nobel lectures, physics 1991–1995* (pp. 168–195). Singapore: World Scientific Publishing Co. (Nobel Prize acceptance speech delivered on December 8, 1995).

Perl, M.L. (2004). The discovery of the Tau Lepton and the changes in elementary-particle physics in forty years. *Physics in Perspective, 6*, 401–427.

Perl, M.L. (2007). A contrarian view of how to develop creativity in science and engineering. Paper presented at The Eighth Olympiad of the Mind, The National Academies, Washington, DC, November (SLAC-PUB-12850).

Perl, M.L., & Lee, E.R. (1997). The search for elementary particles with fractional electrical charge and the philosophy of speculative experiments. *American Journal of Physics, 65*, 698–706.

Perl, M.L., Lee, E.R., & Loomba, D. (2004). A brief review of the search for isolatable fractional charge elementary particles. *Modern Physics Letters, A19*, 2595–2610.

Perl, M.L., Abrams, G.S., Boyarski, A.M., Breidenbach, M., Briggs, D.D., Bulos, F., et al. (1975). Evidence for anomalous lepton production in e^+–e^- annihilation. *Physics Review Letters, 35*, 1489–1492.

Perry, W.G. (1999). *Forms of ethical and intellectual development in the college years: A scheme*. San Francisco: Jossey-Bass.

Petrucci, D., & Dibar, M.C. (2001). Imagen de la ciencia en alumnos universitarios: Una revisión y resultados. *Enseñanza de las Ciencias, 19*, 217–229.

Phillips, D.C. (2005). The contested nature of scientific educational research: A guide for the perplexed. Keynote address at the 11th Biennial European Conference for Research on Learning and Instruction, Nicosia, Cyprus (available online: http://earli2005conference.ac.cy).

Phillips, D.C., & Burbules, N.C. (2000). *Postpositivism and educational research.* New York: Rowman & Littlefield.

Pintrich, P.R., Marx, R.W., & Boyle, R.A. (1993). Beyond cold conceptual change: The role of motivation beliefs and classroom contextual factors in the process of conceptual change. *Review of Educational Research, 63*, 167–199.

PLUTO–Collaboration,et al. (1977). Anomalous muon production e^+e^- annihilations as evidence for heavy leptons. *Physics Letters B, 68*, 297–300.

Pocoví, M.C. (2007). The effects of a history–basedinstructional material on the students' understanding of field lines. *Journal of Research in Science Teaching, 44*, 107–132.

Polanyi, M. (1964). *Personal knowledge: Towards a post-critical philosophy.* Chicago: University of Chicago Press (first published 1958).

Polanyi, M. (1967). *The tacit dimension.* London: Routledge & Kegan Paul.

Popper, K. (1965). *Conjectures and refutations. The growth of scientific knowledge.* New York: Basic Books.

Posner, G.J., Strike, K.A., Hewson, P.W., & Gertzog, W.A. (1982). Accommodation of a scientific conception: Towards a theory of conceptual change. *Science Education, 67*, 489–508.

Raman, V.V., & Forman, P. (1969). Why was it Schrödinger who developed de Broglie's ideas? *Historical Studies in the Physical Sciences, 1*, 291–314.

Reiner, M., & Gilbert, J.K. (2004). The symbiotic roles of empirical experimentation and thought experimentation in the learning of physics. *International Journal of Science Education, 26*, 1819–1834.

Reines, F. (1997). The neutrino: From poltergeist to particle. In G. Ekspong (Ed.), *Nobel lectures, physics, 1991–1995* (pp. 202–221). Singapore: World Scientific Publishing Co. (Nobel Prize acceptance speech delivered on December 8, 1995).

Robinson, J.T. (1969a). Philosophy of science: Implications for teacher education. *Journal of Research in Science Teaching, 6*, 99–104.

Robinson, J.T. (1969b). *The nature of science and science teaching.* Belmont, CA: Wadsworth.

Rocke, A.J. (1978). Atoms and equivalents: The early development of the chemical atomic theory. *Historical Studies in the Physical Sciences, 9*, 225–263.

Rocke, A.J. (1984). *Chemical atomism in the nineteenth century: From Dalton to Cannizaro.* Columbus: Ohio State University Press.

Rodríguez, M.A., & Niaz, M. (2002). How in spite of the rhetoric, history of chemistry has been ignored in presenting atomic structure in textbooks. *Science & Education, 11*, 423–441.

Rodríguez, M.A., & Niaz, M. (2004a). La teoría cinético–moleculardе los gases en libros de fisica: Una perspectiva basada en la historia y filosofia de la ciencia. *Revista de Educación en Ciencias, 5*, 68–72.

Rodríguez, M.A., & Niaz, M. (2004b). A reconstruction of structure of the atom and its implications for general physics textbooks. *Journal of Science Education and Technology, 13*, 409–424.

Rodríguez, M.A., & Niaz, M. (2004c). The oil drop experiment: An illustration of scientific research methodology and its implications for physics textbooks. *Instructional Science, 32*, 357–386.

Roehrig, G.H., & Luft, J.A. (2004). Research report. *International Journal of Science Education, 26*, 3–24.

Rosenfeld, L. (1963). Introduction to Bohr's *On the constitution of atoms and molecules* (pp. xliv–xlv). Copenhagen: Munksgaard.

Rowell, J.A., & Cawthron, E.R. (1982). Images of science: An empirical study. *European Journal of Science Education, 4*, 79–94.
Rudolph, J.L. (2008). Historical writing on science education: A view of the landscape. *Studies in Science Education, 44*, 63–82.
Russ, R.S., Scherr, R.E., Hammer, D., & Mikeska, J. (2008). Recognizing mechanistic reasoning in student scientific inquiry: A framework for discourse analysis developed from philosophy of science. *Science Education, 92*, 499–525.
Russell, C.A. (1988). Rude and disgraceful beginnings: A view of history of chemistry from the nineteenth century. *British Journal for the History of Science, 21*, 273–294.
Rutherford, E. (1911). The scattering of alpha and beta particles by matter and the structure of the atom. *Philosophical Magazine, 21*, 669–688.
Ryder, J., Leach, J., & Driver, R. (1999). Understanding science students' images of science. *Journal of Research in Science Teaching, 36*, 201–219.
Sadler, T.D. (2004). Informal reasoning regarding socioscientific issues: A critical review of research. *Journal of Research in Science Teaching, 41*, 513–536.
Sadler, T.D., Chambers, F.W., & Zeidler, D.L. (2004). Student conceptualizations of the nature of science in response to a socio-scientific issue. *International Journal of Science Education, 26*, 387–409.
Sadler, T.D., & Zeidler, D.L. (2005). Patterns of informal reasoning in the context of socioscientific decision making. *Journal of Research in Science Teaching, 42*, 112–138.
Samarapungavan, A., Westby, E.L., & Bodner, G.M. (2006). Contextual epistemic development in science: A comparison of chemistry students and research chemists. *Science Education, 90*, 468–495.
Sampson, V., & Clark, D.B. (2008). Assessment of the ways students generate arguments in science education: Current perspectives and recommendations for future directions. *Science Education, 92*, 447–472.
Sandoval, W.A. (2005). Understanding students' practical epistemologies and their influence on learning through inquiry. *Science Education, 89*, 634–656.
Sandoval, W.A., & Reiser, B.J. (2004). Explanation-driven inquiry: Integrating conceptual and epistemic scaffolds for scientific inquiry. *Science Education, 88*, 345–372.
Scharmann, L.C., & Smith, M.U. (2001). Further thoughts on defining versus describing the nature of science: A response to Niaz. *Science Education, 85*, 691–693.
Scheffler, I. (1992). Philosophy and the curriculum. *Science and Education, 1*(4), 385–394.
Schrödinger, E. (1922). Bemerkenswerte eigenschaft der quantenbahnen? *Zeitschrift für Physik, 12*, 13–23.
Schrödinger, E. (1926a). On Einstein's gas theory. *Physikalische Zeitschrift, 27*, 95.
Schrödinger, E. (1926b). *Annalen de Physik, 79*, 734.
Schulz, R.M. (2009a). Reforming science education: Part I. The search for a *philosophy* of science education. *Science & Education, 18*, 225–249.
Schulz, R.M. (2009b). Reforming science education: Part II. Utilizing Kieran Egan's educational metatheory. *Science & Education, 18*, 251–273.
Schwab, J.J. (1962). *The teaching of science as enquiry*. Cambridge, MA: Harvard University Press.
Schwab, J.J. (1964). Problems, topics, and issues. In S. Elam (Ed.), *Education and the structure of knowledge* (pp. 4–43). Chicago: Rand McNally & Company.
Schwab, J.J. (1974). The concept of the structure of a discipline. In E.W. Eisner & E. Vallance (Eds.), *Conflicting conceptions of curriculum* (pp. 162–175). Berkeley, CA: McCutchan Publishing Corp.

Schwartz, R.S., & Lederman, N.G. (2008). What scientists say: Scientists' views of nature of science and relation to science context. *International Journal of Science Education, 30*, 727–771.

Schwartz, R.S., Lederman, N.G., & Crawford, B.A. (2004). Developing views of nature of science in an authentic context: An explicit approach to bridging the gap between nature of science and scientific inquiry. *Science Education, 88*, 610–645.

Segal, B.G. (1989). *Chemistry: Experiment and theory* (2nd ed.). New York: Wiley.

Sensevy, G., Tiberghein, A., Santini, J., Laubé, S., & Griggs, P. (2008). An epistemological approach to modeling: Case studies and implications for science teaching. *Science Education, 92*, 424–446.

Sevian, H., & Gonsalves, L. (2008). Analysing how scientists explain their research: A rubric for measuring the effectiveness of scientific explanations. *International Journal of Science Education, 30*(11), 1441–1467.

Shapere, D. (1977). Scientific theories and their domains. In F. Suppe (Ed.), *The structure of scientific theories* (2nd ed., pp. 518–565). Chicago: University of Illinois Press.

Shapin, S. (1996). *The scientific revolution*. Chicago: University of Chicago Press.

Shiland, T.W. (1997). Quantum mechanics and conceptual change in high school chemistry textbooks. *Journal of Research in Science Teaching, 34*, 535–545.

Shipman, H.L. (1980). *Black holes, quasars, and the universe*. Boston: Houghton Mifflin.

Shulman, L.S. (1986). Paradigms and research programs in the study of teaching: A contemporary perspective. In M.C. Wittrock (Ed.), *Handbook of research on teaching* (3rd ed., pp. 3–36). New York: Macmillan.

Silberberg, M.S. (2000). *Chemistry: The molecular nature of matter and change* (2nd ed.). New York: McGraw-Hill.

Simon, S., Erduran, S., & Osborne, J. (2006). Learning to teach argumentation: Research and development in the science classroom. *International Journal of Science Education, 28*, 235–260.

Simon, S., Osborne, J., & Erduran, S. (2003). Systematic teacher development to enhance the use of argumentation in school science activities. In J. Wallace & J. Loughran (Eds.), *Leadership and professional development in science education* (pp. 198–217). London: RoutledgeFalmer.

Sisler, H.H., Dresdner, R.D., & Mooney, W.T. (1980). *Chemistry: A systematic approach*. New York: Oxford University Press.

Skoog, G. (2005). The coverage of human evolution in high school biology textbooks in the 20th century and in current state science standards. *Science & Education, 14*, 395–422.

Slater, M. (2008). How to justify teaching false science. *Science Education, 92*, 526–542.

Smart, W.M. (1946). John Couch Adams and the discovery of Neptune. *Nature, 158*, 648–652.

Smith, C.L., & Wenk, L. (2006). Relations among three aspects of first-yearcollege students' epistemologies of science. *Journal of Research in Science Teaching, 43*, 747–785.

Smith, M.U., Lederman, N.G., Bell, R.L., McComas, W.F., & Clough, M.P. (1997). How great is the disagreement about the nature of science? A response to Alters. *Journal of Research in Science Teaching, 34*, 1101–1103.

Smith, M.U., & Scharmann, L.C. (1999). Defining versus describing the nature of science: A pragmatic analysis for classroom teachers and science educators. *Science Education, 83*, 493–509.

Smith, M.U., & Scharmann, L.C. (2008). A multi-yearprogram developing an explicit reflective pedagogy for teaching pre-serviceteachers the nature of science by ostention. *Science & Education, 17*, 219–248.

Sodian, B., Zaitchik, D., & Carey, S. (1991). Young children's differentiation of hypothetical beliefs from evidence. *Child Development, 62*, 753–766.

Solomon, J. (1998). About argument and discussion. *School Science Review, 80*, 57–62.

Solomon, J., Scott, L., & Duveen, J. (1996). Large-scaleexploration of pupils' understanding of the nature of science. *Science Education, 80*, 493–508.

Sommerfeld, A. (1916). *Annalen der Physik, 51*, 1–94, 125–167.

Southerland, S.A., Gess-Newsome, J., & Johnston, A. (2003). Portraying science in the classroom: The manifestation of scientists' beliefs in classroom practice. *Journal of Research in Science Teaching, 40*, 669–691.

Southerland, S.A., Johnston, A., & Sowell, S. (2006). Describing teachers' conceptual ecologies for the nature of science. *Science Education, 90*, 874–906.

Sowell, S., Johnston, A., & Southerland, S. (2007). Calling for a focus on where learning happens: A response to Abd-El-Khalickand Akerson. *Science Education, 91*, 195–199.

Staver, J.R. (1998). Constructivism: Sound theory for explicating the practice of science and science teaching. *Journal of Research in Science Teaching, 35*, 501–520.

Stern, L., & Roseman, J.E. (2004). Can middle-schoolscience textbooks help students learn important ideas? Findings from Project 2061's curriculum evaluation study: Life science. *Journal of Research in Science Teaching, 41*, 538–568.

Stinner, A. (1992). Science textbooks and science teaching: From logic to evidence. *Science Education, 76*, 1–16.

Stinner, A. (1995). Contextual settings, science stories, and large context problems: Toward a more humanistic science education. *Science Education, 79*, 555–581.

Stinner, A., McMillan, B., Metz, D., Jilek, J., & Klassen, S. (2003). The renewal of case studies in science education. *Science & Education, 12*(7), 617–643.

Stoker, H.S. (1990). *Introduction to chemical principles* (3rd ed.). New York: Macmillan.

Strike, K.A., & Posner, G.J. (1992). A revisionist theory of conceptual change. In R.A. Duschl & R.J. Hamilton (Eds.), *Philosophy of science, cognitive psychology, and educational theory and practice*. Albany: State University of New York Press.

Stuewer, R.H. (1975). *The Compton effect: Turning point in physics*. New York: Science History Publications.

Tampakis, C., & Skordoulis, C. (2007). The history of teaching quantum mechanics in Greece. *Science & Education, 16*, 371–391.

Taylor, A.R., Gail Jones, M., Broadwell, B., & Oppewal, T. (2008). Creativity, inquiry, or accountability? Scientists' and teachers' perceptions of science education. *Science Education, 92*, 1058–1075.

Taylor, H.S. (1942). The atomic concept of matter. In H.S. Taylor & S. Glasstone (Eds.), *A treatise on physical chemistry*, Vol. 1. Princeton, NJ: D. van Nostrand.

Thomson, J.J. (1897). Cathode rays. *Philosophical Magazine, 44*, 293–316.

Thomson, J.J. (1907). *The corpuscular theory of matter*. London: Constable.

Thomson, J.J. (Ed.). (1918). *Natural science in education*. London: HMSO (known as the Thomson Report).

Thomson, T. (1825). *An attempt to establish the first principles of chemistry by experiments* (2 vols.). London: Colburn & Bentley.

Toulmin, S. (1958). *The uses of argument*. Cambridge, UK: Cambridge University Press.

Trend, R.D. (1998). An investigation into understanding of geological time among 10-and 11-yearold children. *International Journal of Science Education, 20*(8), 973–988.

Trend, R.D. (2000). Conceptions of geological time among primary teacher trainees with reference to their engagement with geoscience, history and science. *International Journal of Science Education, 22*(5), 539–555.

Trend, R.D. (2001). Deep time framework: A preliminary study of U.K. primary teachers' conceptions of geological time and perceptions of geoscience. *Journal of Research in Science Teaching, 38*, 191–221.

Tsai, C.-C. (1998). An analysis of scientific epistemological beliefs and learning orientations of Taiwanese eighth graders. *Science Education, 82*, 473–489.

Tsai, C.-C.(1999). "Laboratory exercises help me memorize the scientific truths": A study of eighth graders' scientific epistemological views and learning in laboratory activities. *Science Education, 83*, 654–674.

Tsai, C.-C. (2006). Reinterpreting and reconstructing science: Teachers' view changes toward the nature of science by courses of science education. *Teaching and Teacher Education, 22*, 363–375.

Tsai, C.-C.(2007). Teachers' scientific epistemological views: The coherence with instruction and students' views. *Science Education, 91*, 222–243.

Tsai, C.-C.,& Liu, S.-Y.(2005). Developing a multi-dimensionalinstrument for assessing students' epistemological views toward science. *International Journal of Science Education, 27*, 1621–1638.

Tsai, C.-C.,& Wen, M.L. (2005). Research and trends in science education from 1998 to 2002: A content analysis of publication in selected journals. *International Journal of Science Education, 27*, 3–14.

Tytler, R., & Peterson, S. (2004). From "try it and see" to strategic exploration: Characterizing young children's scientific reasoning. *Journal of Research in Science Teaching, 41*, 94–118.

Urone, P.P. (2001). *College physics* (2nd ed.). Pacific Grove, CA: Brooks/Cole.

Van Aalsvoort, J. (2004). Logical positivism as a tool to analyse the problem of chemistry's lack of relevance in secondary school chemical education. *International Journal of Science Education, 26*, 1151–1168.

Van Berkel, B., DeVos, W., Verdonk, A.H., & Pilot, A. (2000). Normal science education and its dangers: The case of school chemistry. *Science & Education, 9*, 123–159.

Van der Waals, J.D. (1873). *Over de continuiteit van den gas en vloeistoftoestand*. Leyden: University of Leyden.

Van Spronsen, J. (1969). *The periodic system of chemical elements. A history of the first hundred years*. Amsterdam: Elsevier.

Vargas Llosa, M. (2010). Nobel Prize in Literature acceptance speech (http://www.nobelprize.org/nobel_prizes/literature/laureates/2010/vargas_llosa-lecture, downloaded December 11, 2010).

Velentzas, A., Halkia, K., & Skordoulis, C. (2007). Thought experiments in the theory of relativity and in quantum mechanics: Their presence in textbooks and in popular science books. *Science & Education, 16*, 353–370.

Vhurumuku, E., Holtman, L., Mikalsen, O., & Kolstø, S. (2006). An investigation of Zimbabwe high school chemistry students' laboratory work-basedimages of nature of science. *Journal of Research in Science Teaching, 43*, 127–149.

Villani, A. (1992). Conceptual change in science and science education. *Science Education, 76*, 223–237.

Volkmann, M.J., & Zgagacz, M. (2004). Learning to teach physics through inquiry: The lived experience of a graduate teaching assistant. *Journal of Research in Science Teaching, 41*, 584–602.

Von Aufschnaiter, C., Erduran, S., Osborne, J., & Simon, S. (2008). Arguing to learn and learning to argue: Case studies of how students' argumentation relates to their scientific knowledge. *Journal of Research in Science Teaching, 45*, 101–131.

Wartofsky, M.W. (1968). *Conceptual foundations of scientific thought: An introduction to the philosophy of science.* New York: Macmillan.

Warwick, P., & Siraj–Blatchford, J. (2006). Using data comparison and interpretation to develop procedural understandings in the primary classroom: Case study evidence from action research. *International Journal of Science Education, 28*, 443–467.

Weinberg, S. (2001). Physics and history. In J.A. Labinger & H.M. Collins (Eds.), *The one culture: A conversation about science.* Chicago: University of Chicago Press.

Weisberg, M. (2007). Who is a modeler? *British Journal for the Philosophy of Science, 58*, 207–233.

Westerlund, J., & Fairbanks, D. (2004). Gregor Mendel and "myth conceptions." *Science Education, 88*, 754–758.

Westfall, R.S. (1986). *The construction of modern science: Mechanisms and mechanics.* New York: Cambridge University Press.

Wheaton, B.R. (1983). *The tiger and the shark: Empirical roots of wave-particle dualism.* Cambridge: Cambridge University Press.

Wiebe, R., & Stinner, A. (2010). Using story to help student understanding of gas behavior. *Interchange, 41*(4), 347–361.

Wiechert, E. (1897). Ergebniss einer messung der geschwindigkeit der cathodenstrahlen. *Schriften der Physikalischokonomisch Gesellschaft zu Konigsberg, 38*, 3.

Wilson, D. (1983). *Rutherford: Simple genius.* Cambridge, MA: MIT Press.

Winchester, I. (2006). Beyond the textbook: Reaching out to the context and to history. *Interchange, 37*, 1–2.

Windschitl, M. (2004). Folk theories of "inquiry": How preservice teachers reproduce the discourse and practices of an atheoretical scientific method. *Journal of Research in Science Teaching, 41*, 481–512.

Windschitl, M., Thompson, J., & Braaten, M. (2008). Beyond the scientific method: Model–basedinquiry as a new paradigm of preference for school science investigations. *Science Education, 92*, 941–967.

Wiser, M. (1988). The differentiation of heat and temperature: History of science and novice–expertshift. In S. Strauss (Ed.), *Ontogeny, phylogeny and historical development* (pp. 28–48). Norwood, NJ: Ablex.

Wong, S.L., Hodson, D., Kwan, J., Wai Jung, B.H. (2008). Turning crisis into opportunity: Enhancing student–teachers' understanding of nature of science and scientific inquiry through a case study of the scientific research in severe acute respiratory syndrome. *International Journal of Science Education, 30*(11), 1417–1439.

Yalvac, B., Tekkaya, C., Cakiroglu, J., & Kahyaoglu, E. (2007). Turkish pre–servicescience teachers' views on science–technology–societyissues. *International Journal of Science Education, 29*, 331–348.

Yang, F.–Y. (2004). Exploring high school students' use of theory and evidence in an everyday context: The role of scientific thinking in environmental science decision–making. *International Journal of Science Education, 26*, 1345–1364.

Yavuz, O. (1978). *Genel kimya.* Erzurum, Turkey: Atatürk Üniversitesi Basimevi.

Yilmaz–Tuzun, O., & Topcu, M.S. (2008). Relationships among preservice science

teachers' epistemological beliefs, epistemological world views, and self–efficacybeliefs. *International Journal of Science Education, 30*, 65–85.
Yore, L.D., Florence, M.K., Pearson, T.W., & Weaver, A.J. (2006). Written discourse in scientific communities: A conversation with two scientists about their views of science, use of language, role of writing in doing science, and compatibility between their epistemic views and language. *International Journal of Science Education, 28*, 109–141.
Yore, L.D., Hand, B.M., & Florence, M.K. (2004). Scientists' views of science, models of writing, and science writing practices. *Journal of Research in Science Teaching, 41*, 338–369.
Ziman, J. (1978). *Reliable knowledge. An exploration of the grounds for belief in science.* Cambridge, UK: Cambridge University Press.
Zohar, A., & Aharon–Kravetsky,S. (2005). Exploring the effects of cognitive conflict and direct teaching for students of different academic levels. *Journal of Research in Science Teaching, 42*, 829–855.
Zohar, A., & Nemet, F. (2002). Fostering students' knowledge and argumentation skills through dilemmas in human genetics. *Journal of Research in Science Teaching, 39*, 35–62.
Zumdahl, S.S. (1990). *Introductory chemistry: A foundation.* Lexington, MA: Heath.

INDEX